FRESH WATER SEAS

FRESH WATER SEAS

saving the great lakes

PHIL WELLER

between the lines

Published by Between The Lines
394 Euclid Avenue, #203
Toronto, Ontario
M6G 2S9
Canada

Cover design by Linda Gustafson Design
Typeset by The Coach House Press, Toronto
Printed in Canada

Between The Lines receives financial assistance from the Ontario Arts Council, the Canada Council, and the Department of Communications.

The map on p. 10 is used with the permission of the International Joint Commission and the Great Lakes Water Quality Board.

CANADIAN CATALOGUING IN PUBLICATION DATA

Weller, Phil, 1956-
Fresh water seas: saving the Great Lakes

ISBN 0-921284-18-7 (bound) ISBN 0-921284-19-5 (pbk.)

1. Great Lakes – History. 2. Water – Pollution – Great Lakes. 3. Lake ecology – Great Lakes. I. Title.

TD223.3.W44 1990 363.73'94'0977 C90-094804-3

Printed on paper
containing over 50%
recycled paper including
5% post-consumer fibre.

Contents

PART THREE

Acknowledgements

THE IDEA FOR this book was born in 1985. At the time I was a partner in a small, successful company that combined environmental activism with research, writing, and consulting on environmental issues. The challenge of the business was to make a living as well as doing work we wanted to do. One of the projects that arose from that partnership was a book on the Great Lakes that would provide a context for today's problems and at the same time reflect the challenges, hopes, and aspirations of those who are struggling to respond to the environmental problems in the Great Lakes region.

While ideas for projects and books are sometimes easy to come by, money and support for doing them are not. This book would not have gotten past the dream stage were it not for the interest, encouragement, and financial support given by Henry Regier and George Francis. A significant portion of my appreciation and knowledge of the lakes can be traced to my association and friendship with them. Together with their colleagues in the Great Lakes Ecosystem Rehabilitation Network (GLER), they funded the research and writing of this book and provided critiques and reviews at critical points. Most of all, their never-ending work on behalf of the Great Lakes was an essential inspiration. Their funding was made possible through a grant from the Donner Canadian Foundation.

Other members of the GLER Network – Sally Lerner, Tom Whillans, Lino Grima, and Fikret Berkes – provided important assistance and ideas. Tom Whillans furnished an especially detailed critique and review. Sally Lerner offered not only a thorough critique but also encouragement and support when spirits were lagging.

Throughout the course of conducting the research for this book I benefited from generous gifts of knowledge and insight from a considerable number of people. I would especially like to thank Jack Vallentyne, Lee Botts, Don Misner, Harry Barrett, Bud Harris, Becky Leighton, Rich Thomas, John Gannon, Ron Shimizu, Rick Findlay, Sister Margeen Hoffman, Tim Eder, Dave Miller, and John Black.

Pat Murray and the staff of the International Joint Commission Library in Windsor were extremely helpful. Raf Serafin provided a valuable review of an early version of the manuscript and enjoyable and stimulating company on research trips. Jill Singer helped fill me in on the geological history of the Great Lakes. Grant Jarvis contributed advice on the section dealing with Native history. In addition, his friendship, encouragement, and support were invaluable.

Jamie Swift and others at Between The Lines provided detailed editorial comments. Robert Clarke is owed a particularly large debt of thanks for the patient and thorough manner in which he helped fill out and polish the finished manuscript. Words Work at the University of Waterloo assisted with an earlier version of the manuscript.

Special thanks go to Don Sedgwick, who generously advised and assisted in the early stages of developing the book. Jim Weller, my father, provided editorial advice and created the index. The Ontario Arts Council supplied much needed financial assistance through its Writers Reserve Program, and a large number of friends and family members gave support and encouragement, without which this book would not have been possible.

Acknowledgements are not complete without the author accepting full responsibility for any inaccuracies or omissions, and of course I do. The acknowledgements, however, would truly not be complete without giving special thanks for inspiration to all those individuals and organizations who have worked to ensure that the environmental health of our Great Lakes home is protected and restored.

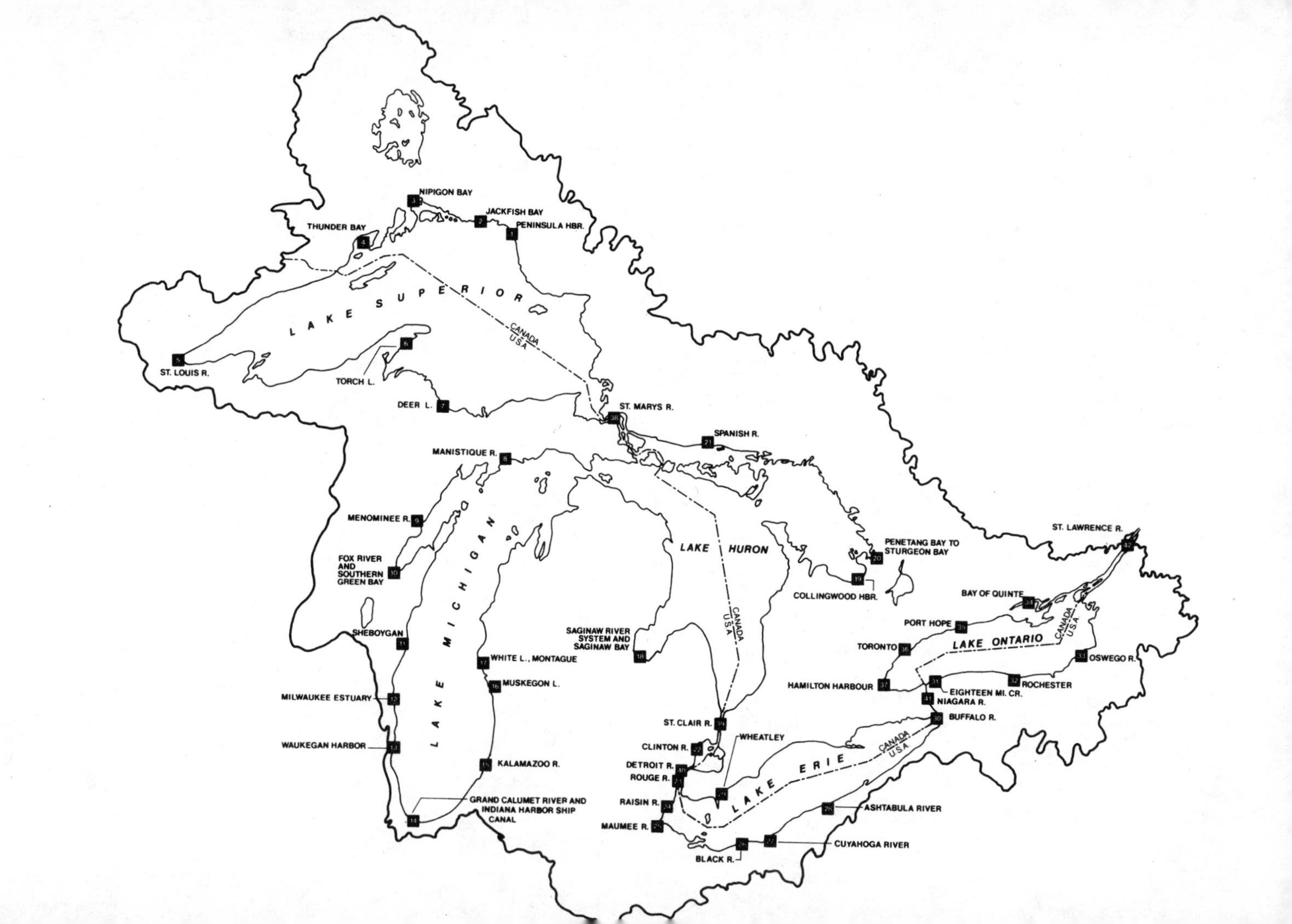
LAKE SUPERIOR
LAKE MICHIGAN
LAKE HURON
LAKE ONTARIO
LAKE ERIE
CANADA
U.S.A.
NIPIGON BAY
JACKFISH BAY
PENINSULA HBR.
THUNDER BAY
ST. LOUIS R.
TORCH L.
DEER L.
ST. MARYS R.
SPANISH R.
MANISTIQUE R.
MENOMINEE R.
FOX RIVER AND SOUTHERN GREEN BAY
SHEBOYGAN
MILWAUKEE ESTUARY
WAUKEGAN HARBOR
GRAND CALUMET RIVER AND INDIANA HARBOR SHIP CANAL
KALAMAZOO R.
MUSKEGON L.
WHITE L., MONTAGUE
SAGINAW RIVER SYSTEM AND SAGINAW BAY
PENETANG BAY TO STURGEON BAY
COLLINGWOOD HBR.
ST. LAWRENCE R.
BAY OF QUINTE
PORT HOPE
TORONTO
HAMILTON HARBOUR
EIGHTEEN MI. CR.
NIAGARA R.
ROCHESTER
OSWEGO R.
BUFFALO R.
ST. CLAIR R.
CLINTON R.
WHEATLEY
DETROIT R.
ROUGE R.
RAISIN R.
MAUMEE R.
BLACK R.
CUYAHOGA RIVER
ASHTABULA RIVER

INTRODUCTION

A Region in Crisis

THE VISIBLE DECAY of the Great Lakes environment had a profound effect on my childhood. As a boy growing up in Toronto, I used to fish, with little chance of success, in the murky brown waters of the Don River. Other times, from the deck of the Toronto Island ferry, I watched the bodies of small dead fish float by in the multicoloured water of Toronto harbour. From my playground in the Don River valley, a deep and wide north-south trough that cuts Toronto in half, I listened and watched as bulldozers cleared trees and fields to make way for townhouses needed by the expanding population of the city.

These images from my youth often return to me, but never more clearly than one spring day in the late 1980s when I sat on the shore of Lake Ontario near the mouth of the Niagara River. To the north I could clearly see the Toronto skyline standing out on the horizon. The deep blue water of the lake before me was dotted with the white and multicoloured sails of windsurfers and yachts. The faint smell of rotting algae rose from the rocks below where children were skipping stones out onto the low waves. It was a picture-postcard scene, yet I was filled with a sense of things gone wrong.

For over two years I had been consumed with the task of researching the ecological history of the Great Lakes region. The shelves of my office had become heavy with reports on the pollution of the lakes, the decline of the fish, the scars on the land. But reports and books were not all I had collected. My trips throughout the region had added experiences and images that reinforced the painful lesson I had learned early in life – that the cumulative progress of our industrial and profit-based society had dramatically undermined the ecological health of the Great Lakes environment.

In the summer of 1987 I had explored the shores of Green Bay, Wisconsin, with ecologist Bud Harris, who turned to me and said, "The land you are now standing on was once one of the largest and most productive marshes in the Great Lakes." Looking around I saw only dry earth and rubble, into which a sign advertising industrial lots had been driven. A few months earlier I had gone to Buffalo and Niagara Falls to conduct interviews for this book. At Buffalo's Roswell Park Memorial Laboratory I peered through a microscope at cancerous fish tumors, which are thought to be caused at least partially by exposure to industrial chemicals. Later that day I sat in the office of the Ecumenical Taskforce in Niagara Falls and heard the strong but controlled anger of Sister Margeen Hoffman as she described the shortsighted approach to waste management along the Niagara River. As Sister Margeen talked, Pat Brown, a victim of the horrors of Love Canal, sat alongside her, a witness to the insensitivity and destructiveness of industrial activity.

Listening to them, I wanted to believe that their stories were simply aberrations, a matter of isolated mistakes. But from other interviews and experiences I knew otherwise. I remembered a trip to Tiny Township, a rural area in Ontario to the east of Georgian Bay. Walking around a huge pile of garbage, Jean Terrien, a quiet local woman, had told her tale of exposure to chemical contamination: hazardous liquid industrial waste had been carelessly dumped into a nearby landfill. A few months after I visited her Jean Terrien ended up in the district hospital, suffering from a disease that may well have been caused or aggravated by exposure to the large amounts of Trichlorethylene in her water supply.

Then there was the time, on a canoe trip into the interior of Ontario's Algonquin Park, when I stared up at one of the last remaining stands of virgin white pine in the Great Lakes region. The twelve tall pines were a pitiful reminder of the area's once great forests.

The damage that human activity has done to the Great Lakes environment over the past century is a disgrace. The lakes are now home to about one thousand different chemicals. Every day some

3,630 kilograms – or 8,000 pounds – of toxic chemicals enter the lakes, the nearby land, and air. Day by day thousands of tons of buried wastes, including some of the most deadly substances known, leak slowly into the land around the lakes. We have hacked down the original forests, destroyed important wetlands, and made other major alterations to the landscape. The accidental introduction of the lamprey, the elimination of the Atlantic salmon from Lake Ontario, and the complete extinction of the blue pike: these kinds of changes have seriously undermined the quality of all life in the region.

The abuses of the lakes and the surrounding landscape are many and varied, and no single factor has led to the area's deterioration. Rather, it has been the long-time cumulative abuses of the environment that have brought the basin to its present condition. The Great Lakes *ecosystem* – which includes all the interacting components of air, water, minerals, and living organisms within the drainage area of the Great Lakes – is complex and always changing. In the past two hundred years, however, the patterns of economic and social development, along with our failure to recognize ourselves as part of this ecosystem, have caused rapid and destructive changes, fundamentally undermining the health of the region.

But the decline of the Great Lakes environment is not the inevitable consequence of human activity. Within recent decades I have seen encouraging signs that we can put a halt to the deterioration and destruction. Individuals and groups have made determined efforts to stop the abuses, to preserve the remaining attributes of the Great Lakes environment, and to restore the lakes to a condition of health. These efforts are bringing new hope that the region will once again be worthy of the description "Great."

The road to restoring health is not easy. It is not enough simply to stop a percentage of the abuses. Winning only some of the necessary battles will merely prolong the gradual drift of the Great Lakes region towards ecological collapse. The damage continues to be done and there is always resistance to the small changes that have been achieved. In short, we must win not only individual battles but also the whole war against abuse of the Great Lakes environment. We must learn to understand how this region works and grows and we must begin to implement the strategies necessary to restore and rebuild the degraded parts of its environment.

PART ONE

ONE

The Evolution of an Ecosystem

AS WE FLY NORTH from Washington, D.C., en route to Toronto via Rochester, the sky is clear, with an occasional small cloud illuminated by the low sun. We have been on the plane for almost an hour and my feeling of tension from the past two days of meetings has begun to disappear. Out the window, towards the east, I can see the Finger Lakes of New York State – they look true to their name. The view of the land from 25,000 feet above is superb.

When we get close to our first stop, in Rochester, the plane turns slightly west and the blinding light of the sun comes into the cabin. Beginning its descent, the plane levels off and for the first time I can see the shoreline of Lake Ontario. The sun shining on the lake makes it look like a sheet of glass. Then I see the city, and notice a large inlet along the shore, guarded by what appears to be a large sandbar. Small boats crowd the lakefront. From the sky, Rochester looks clean and attractive, with its tiny cars and streets and neatly spaced downtown buildings. Everything in sight now seems to be the product of human hands or machinery.

A few minutes later the plane takes off again, this time for Toronto, and we can see the southern shoreline of the lake. Climbing quickly, we can still make out the white dots of sailboats on the water, and speed boats too, each one with a trailing tail of foam. Then, as we begin our descent, the Canadian shore of the lake emerges in the distance. Soon we can clearly see the Toronto waterfront and the long arms of the Leslie Street spit, an artificial peninsula. The Toronto islands, which protect the harbour, are a rich green from the summer growth of trees.

The plane flies past one of the newest creations, the huge white dome of the baseball stadium, strategically located right next to the gigantic spire of the CN Tower. Then we can see the smaller geodesic dome of Ontario Place, a recreational park on the waterfront. As the plane arches west past the mouths of the small Humber and Credit rivers, the great number of sailboats moored along the shore is a reminder of how the lake is used by many of the city residents. Then the plane moves north, away from the lake, and the land below becomes an endless expanse of roads, houses, and shopping malls, interrupted here and there by the exposed earth of a new construction site. I am struck by how little wooded green space there is in this part of the city. The immensity of Toronto from the air is astonishing, the extent of land covered over is remarkable.

I'm entranced by this act, of watching the land below. As the plane slowly makes its way towards the ground, I find myself thinking that the Great Lakes landscape – the water and waterways, the farmland and forests, the rugged land and the tamed land – has very much influenced the character of the people who live there. But the people – their homes, their institutions, and their enterprises – have also greatly influenced the lakes and the surrounding environment. What follows is the story of the Great Lakes region and the impact of humans upon it. And, ultimately, the challenge facing all of us who want to save the region from its own destruction.

A GREAT LAKES HOME: FROM SUPERIOR TO THE ST. LAWRENCE

From the time I arrived in North America at the early age of eighteen months, the Great Lakes region has been my home. I am not alone in this – the region is also home to thirty-seven million other people. The area is immense. It covers some 755,200 square kilometres of land, including the 216,300 square kilometres of the lakes themselves.[1] It takes in eight U.S. states and two Canadian provinces. The territory is urban and suburban, industrial and agricultural, resource-rich and wild.

There is awe-inspiring beauty and grandeur throughout the region. Anyone who has climbed the 140-metre wall of sand at Sleeping Bear Dunes on Lake Michigan, who has stood and marvelled at the massive amounts of water charging over the cataract of Niagara Falls, or who has walked the pebble beach at Agawa Bay in Lake Superior, knows full well the appeal of the region's natural spectacles.

Each of the five main lakes is unique in character and form. Lake Superior, the farthest west and most northerly in the chain, is the largest, coldest, and deepest. At the greatest depth the clear water extends down an amazing 406 metres. Its rugged north shore, composed of the granite rock of the Canadian shield, is heavily forested with spruce, balsam fir, and poplar. The granite hills dip down towards the water, creating wild rocky shores interspersed with quiet peaceful beaches. Towns with evocative names like Red Rock, Marathon, and Terrace Bay hug the coast, nearly all of them depending on the pulp and paper industry for their survival. The rotten-egg smell of the mills warns highway travellers of an upcoming town.

Maples and a variety of other hardwood trees as well as pines cover the more gentle, less rocky southern shore on the U.S. side. Along the southeast edge of the lake, the wind and waves have carved out sandstone cliffs: the spectacular "pictured rocks." Farther west is Duluth, the western terminus of the lakes and, together with the Canadian city of Thunder Bay to the north, a key exchange point in the continental flow of goods in the North American economy. In the harbours of these towns, trains and ships exchange cargoes of wheat and iron.

Lake Superior is an oligotrophic lake; that is, it lacks sufficient nutrients for prolific plant and fish life. Its water is exceptionally clear and the fish that do live in it are ideally suited to the conditions. Flowing southeast, the water of Lake Superior cascades down the seven-metre drop of the St. Marys River into Lake Huron.

Lake Michigan, projecting southward below Lake Superior – and the only one of the Great Lakes to be entirely within the boundaries of a single country – is also nutrient poor, though not as deep as Lake Superior. Green Bay, a 193-kilometre inlet on the northwestern corner, is shallow and more rich in nutrients and produces about half of the fish caught in the lake. Lake Michigan has sand-dune formations along its southern and eastern shores and a scattering of islands in its northern section. The lake's moderating influence on air temperature makes the surrounding landscape ideal for fruit growing, and the orchards around Traverse Bay, on its northeastern corner, produce nearly one-third of the world's cherry crop. Industry also thrives in the Lake Michigan watershed, particularly at the southern end of the lake. The shoreline between the cities of Gary and Chicago is one of the world's most spectacular conglomerations of steel mills and petroleum-tank farms.

The eastward flow of water from Lake Michigan passes under the arches of the Mackinac Bridge to reach Lake Huron, the central link in the chain of five Great Lakes. Lake Huron also receives a slightly larger volume of water from Lake Superior via the St. Marys River, where the twin cities of Sault Ste. Marie are built up along the Michigan and Ontario sides. A massive system of locks raises and lowers boats over the seven-metre drop between the lakes. Before plunging over the St. Marys rapids, the water from Lake Superior passes the piles of coal, limestone, and taconite that feed the Algoma Steel mill.

Lake Huron is also a deep lake – 221 metres at its maximum – and is roughly divided into two sections by the Bruce Peninsula and Manitoulin Island, which separate Georgian Bay and the North Channel from the main body of the lake. The deep water of Georgian Bay is dotted with over twenty thousand islands, providing a summer playground for cottagers and tourists. At the southern end of the lake, the shores are lined with beaches packed with bathers during the hot summer months. The more shallow Saginaw Bay on the western shore of the lake is rich in aquatic life.

Lake Huron water flows south through the narrow channel of the St. Clair River, squeezes under the Bluewater Bridge at Sarnia and Port Huron, and passes the distillation towers of the chemical plants lining the eastern Canadian shore of the river. After a 64-kilometre journey the water empties into shallow Lake St. Clair, which, although large in comparison to many lakes, is not considered one of the five Great Lakes.

Flowing south from Lake St. Clair, the Great Lakes water enters the Detroit River and passes the city of the same name. The city of Windsor on the Canadian shore offers a superb view of the Detroit skyline. South of the Ambassador Bridge, which connects the two cities and countries, the river winds its way past a mass of factories and car plants on its way to Lake Erie.

The water of Lake Erie, the most shallow and southerly of the Great Lakes, has a maximum depth of only 64 metres. The rich, fertile soil of the lake basin is ideal for agriculture, and area farmers grow corn and market vegetables. The U.S. cities of Toledo, Cleveland, Erie, and Buffalo, all on the lakeshore, are centres of industry in contrast to the Canadian shore, where the towns of Port Dover, Wheatley, and Port Stanley are primarily fishing and recreational centres.

Lake Erie, naturally rich in nutrients, is the most biologically productive of all the Great Lakes. Extensive marshes, sand spits, and peninsulas provide a variety of shoreline habitats. From the shallow western end of Lake Erie, the water flows east to the entrance of the Niagara River at the deeper end of the lake. Like all the connecting channels between the lakes, the banks of the Niagara River are lined with industrial plants, especially along the U.S. shore.

After crashing over Niagara Falls, the waters of the four other Great Lakes eventually reach the deep and relatively narrow Lake Ontario. With the smallest surface area of any of the lakes – 18,960 square kilometres – Lake Ontario is, nonetheless, second only to Superior in average depth. At its maximum, Lake Ontario water has a depth of 183 metres. The deep, nutrient-poor water has made it the least productive of the lakes for the fisheries. With the exception of the Bay of Quinte in its northeast corner the shoreline of Lake Ontario is regular and relatively even, and the land that rises slowly from the shores has been extensively converted to agriculture and industry.

In contrast to Lake Erie, the Canadian side of Lake Ontario is the most heavily urbanized and populated. The "Golden Horseshoe," a belt of cities that stretches from Welland in the west to Oshawa in the east and includes Hamilton and Toronto, contains about one-third of the population of Canada. On the U.S. side of the lake, Rochester is the only city of any size along the shoreline, but farther east Syracuse is also within the drainage area of the basin. The last in the Great Lakes chain, Lake Ontario discharges its water into the St. Lawrence River, which flows past the cities of Montreal and Quebec before completing its journey to the Gulf of St. Lawrence and the Atlantic Ocean.

Water is the unifying feature of the region's environment. There is lots of it: 67 trillion gallons in the Great Lakes alone. A freshwater system unsurpassed in magnitude anywhere in the world, these inland seas contain a fifth of the fresh water on the world's surface. The Great Lakes are unique in their vastness; the immense volume of fresh water they hold is a resource of inestimable value. But water is only one of the components of the Great Lakes Basin ecosystem. Soil, plants, animals, fish, humans, and the atmosphere above are all integral parts. The water, the land, the air, and all the living components of the ecosystem interact with and influence one another in a never-ending process of change and evolution.

GLACIAL BEGINNINGS

About three billion years ago a cooling of molten lava formed the solid bedrock that now underlies most of the Great Lakes region.[2] Known as the Canadian Shield, this hard rock is exposed in a number of places around the region, and is especially prominent throughout the area east of Lake Huron towards Ottawa and along the northern shore of Lake Superior.

Other rock layers were added to this foundation. Over five hundred million years ago, shallow seas flooded central North America, creating vast waters inhabited by corals and a host of other marine organisms. The shells and bodies of these organisms formed the basis for the limestone and dolomite rocks that underlie part of the region. The Niagara Escarpment is composed of these rocks; it arches up through the Bruce Peninsula of Ontario and dips below the surface of Lake Huron before resurfacing to form Manitoulin Island, the Michigan upper peninsula, and the Door peninsula on Green Bay. Deposits of mud and sand laid down in the same period were consolidated into shale and sandstone. At the end of this era, known as the Palaeozoic, the seas retreated, leaving behind large rivers that carved out valleys in the less resistant rocks. In later years these valleys widened into the basins that now contain the Great Lakes.

Upon this geologic base, beginning over one hundred million years ago, plant and animal life began to colonize the land and for the next hundred million years evolved and adapted to the conditions of the environment. By two million years ago the climate of the Great Lakes region had slowly begun to cool and the period known as the ice ages had begun. Glaciers crept slowly southward from their centres in northern Canada, with four major periods of retreat and advance. At their greatest extent the glaciers covered all of Canada and the northern United States. At one time the present-day sites of Toronto, Detroit, Cleveland, Milwaukee, and Sault Ste. Marie were all covered with a sheet of ice up to two kilometres thick. The entire Great Lakes region, as far south as the Ohio River, was beneath ice. As the glaciers crept forward they scraped the surface of the rocks and destroyed existing vegetation. Wildlife had to move farther south or perish in the changed conditions. The giant mastodon, which had previously roamed the Great Lakes region, became extinct during the advance and retreat of the ice.

The Great Lakes environment, as we know it, is young by geologic standards. Most of the landforms familiar to us owe their existence to the last ice age, the Wisconsin, which featured a glacier that carved and moulded the landscape. The Wisconsin Glaciation began about seventy-five thousand years ago and reached its maximum extent about fifteen thousand years ago. The scraping action of the glaciers, which followed the existing river valleys, combined with the weight of the ice to form deep depressions in the bedrock of the Great Lakes area.

As the climate warmed, the glacial ice began to melt and retreat northward; large bodies of water, which would eventually evolve into the Great Lakes, began to fill the basins left by the melting glaciers. The glaciers made their final withdrawal from the St. Lawrence Valley about eleven thousand years ago, and meltwaters, which began to form in what is now Lake Ontario, drained to the Atlantic Ocean via the St. Lawrence River. Farther west the huge glacial Lake Algonquin, which would eventually form three of the Great Lakes – Michigan, Huron, and Superior – drained south via the Mississippi River.

This drainage pattern was altered again, about a thousand years later, when the waters of Lake Algonquin started flowing east through what is now Lake Nipissing and the Ottawa Valley to the St. Lawrence River. During this period the water in the basins of lakes Ontario and Erie shrank considerably.

Over the next few thousand years additional drainage outlets developed from what had become three connected upper lakes – Superior, Huron, and Michigan. In addition to the flow of water down the Ottawa River, water flowed south from outlets on lakes Michigan and Huron. Then, about 3000 B.C., the land around North Bay, Ontario, began to rise after being liberated from the weight of the retreating glaciers, a phenomenon called isostatic rebound. This rebounding caused the drainage of the upper lakes to be diverted solely down the St. Clair and Detroit rivers, and the Great Lakes took on the basic shape that we know today.

Lakes and landforms evolve slowly, and the geologic change of the Great Lakes basin has been gradual. Change and evolution continue, even today, to reshape and transform the lakes and the surrounding landscape. The northeastern shore of Lake Superior continues to rise at a rate of up to thirty or fifty centimetres a century as a result of isostatic rebound. Powerful storms occasionally erode large areas of land or lake shoreline, speeding up and graphically illustrating the process of change. Ordinarily, however, change is imperceptible during what, in geologic terms, is merely the instant of a human lifetime.

THE LAND OF GREAT FORESTS AND TROUBLESOME MARSH

When the glaciers made their most recent retreat from the Great Lakes area, plants and animals that had survived to the south started to move northward.[3] By analysing pollen preserved in peat bogs and lake sediments, scientists have pieced together a record of when the various tree species made their way north to the Great Lakes region. Spruce, tamarack, and balsam fir were the first to recolonize the area. Jack pine followed soon after, arriving in the lower lakes about 10,500 to 11,000 years ago. All of these species had made it to the area north of Lake Superior by around 6000 to 7000 B.C. White pine trees increased in abundance in the southern portion of the region about 9000 B.C. and established themselves north of Lake Superior between 6000 and 3000 B.C. Oaks spread northward around the same time and beech trees reached lower Lake Michigan about 5000 B.C.

The resulting forests have never been, and never will be, static and unchanging. The species composition constantly shifts as disease, insects, fire, wind, and climate alter the conditions of growth. The cones of the jackpine, for instance, open up in the heat of a forest fire, a trait making the tree ideally suited for recolonizing burned-over areas. White pine has flexible seed-bed requirements that allow it to germinate in a variety of soil conditions. With their shade tolerance, pine seedlings can survive under a canopy of other trees until an opening in the forest appears. The natural lifetime of pines is 250 to 300 years, and in some places continuous generations of pine have survived for over six thousand years.[4]

When the Europeans arrived they were amazed at the land of forests they found. The tree cover was once described as "so thick-set with trees and so dense" that a native hunter could trek "from Hudson to Lake Erie without exposing himself to the glare of the sun."[5] The majestic white pines, which towered upwards of two hundred feet, were prominent in the forest cover around all of the lakes. Dense upland and swamp forest covered the Lake Erie basin.

Historian Fred Landon described the area that became Michigan as being composed of "great forests." The southern area, according to Landon, "was a part of the hardwood region of the Ohio Valley, having little or no pine or spruce." The northern area was "the home of the evergreens ... On the sandy lands where the pines predominated there were older stands so clear from undergrowth or brush that a team of horses might easily be driven for miles where no road existed."[6]

Throughout southern Ontario a predominantly maple-beech forest evolved. Hemlock, hickory, and elm, a species now almost absent from the landscape because of dutch elm disease, were also common. Huge tracts of white pine grew on sandy soil. Around the shores of southern and western Lake Superior grew red and white pines and magnificent hardwood forests, giving way on the granite north shore to spruce, poplar, and birch.

Still, at the time of the European coming, the forest was not just a continuous unbroken stretch of trees from one end of the basin to the other. Scattered throughout were plenty of beaver meadows and clearings created by fire and wildlife. There were also extensive wetlands. In southern Ontario alone there were some 2.38 million hectares of wetlands, many of them inland swamps and marshes.[7] Hundreds of small bogs were scattered throughout Minnesota on the northwest side of Lake Superior. Large wooded swamps and fens with abundant and beautiful wildflowers, including rare orchids, nestled along the gently sloping shores of the Bruce Peninsula in Ontario. There were vast expanses of marshes on Lake St. Clair and at the shallow western end of Lake Erie. About 150 miles of shoreline between Sandusky and Detroit were covered by marshes where wild rice grew. The largest wetland in the region was probably the Great Black Swamp in northwestern Ohio and eastern Indiana: it covered an amazing 1,500 square miles. In 1791 a Moravian missionary named David Zeisberger described the area as a "troublesome marsh ... where no bit of dry land was to be seen." Horses travelling through the marsh, Zeisberger said, waded up to their knees for miles on end.[8]

The dense cover of vegetation around all the lakes limited soil erosion, so that waters entering the lakes carried low amounts of suspended solids. Biologist Eugene Stoermer concluded that the lakes' waters were originally as "pure as rainwater." This was, he stated, "because the lakes' drainage is small compared to their surface area and because much of the basin is composed of bedrock types that are highly resistant to leaching."[9] With the exception of the shallow basin of Lake Erie, the Great Lakes were all primarily deep, oligotrophic lakes.

Fish distribution in the Great Lakes was greatly influenced by the patterns of drainage that accompanied the withdrawal of the glaciers. Fish such as the muskellunge and the channel catfish entered the region from the south during the period when some of the lake waters drained to southern waterways, such as the Mississippi River. Others found their way into the lakes from the Atlantic Ocean. Atlantic salmon, traditionally a salt-water fish that migrates into freshwater rivers to spawn, established themselves in Ontario when the glacier receded and salt water extended far inland. They subsequently abandoned their traditional migration to the sea and became a freshwater species.

There were also connections between the waters of the Great Lakes and the Arctic, which meant that species such as the Arctic Grayling could enter the Great Lakes and survive near the northern part of the Michigan lower peninsula. Over time the fish of the Great Lakes evolved into a rich mixture of species well adapted to the deep and cool lakes.

Moose came to the region from glacial refuges in the upper Mississippi valley as well as from the Allegheny region of the eastern United States. Deer eventually found their way to southerly areas of the region. Elk, now extinct in eastern North America, roamed the forests along with cougar, lynx, marten, fisher, wolverine, black bear, and timber wolf. The beaver, which provided the mainstay of the fur trade for two hundred years, was everywhere throughout the region. Buffalo herds wandered the prairies of Indiana and Illinois and the unforested areas of Michigan and Wisconsin.

A pair of bald eagles inhabited every five to ten miles of Great Lakes shoreline. Huge flocks of passenger pigeons, consisting of millions of birds, filled the skies. Wild turkeys and large congregations of ducks and geese thrived in the forests of the southern Great Lakes.

Although change was a constant feature of this environment – the composition and abundance of species of plants and animals were forever being altered – the pace of change was slow, dictated by the rhythms of the planet: the time a glacier takes to recede, the time a pine takes to grow and soil to build. The initial human inhabitants of the Great Lakes region were caught up in this never-ending process;

they were part of the ecosystem. They marked the passage of time by the rhythms of nature, by the changing seasons, by the sunrise and sunset, and by the birds' arrival and departure.

A NATIVE ETHIC

Aboriginal peoples lived in the Great Lakes region even as the glaciers disappeared: hunters and gatherers who had migrated to the region from the south.[10] Archeological evidence dating from about 860 B.C. indicates that they hunted caribou and possibly mammoth. Traders from the south introduced corn, or maize, to the region around 500 A.D. and, as a result, some of the native groups began to form settlements in areas suitable for agriculture. In time they also began to grow beans, squash, and sunflower crops.

With the shift to agriculture the native population could expand. By the beginning of the seventeenth century there were about 120,000 native people grouped into several dozen tribes living in the Great Lakes region, with three major linguistic groups. The largest grouping was Iroquoian, who lived around lakes Erie and Ontario, in the St. Lawrence Valley, and near southern Lake Huron. The Algonquian peoples, who included the Ottawa, Montagnais, and Ojibwa, lived in Michigan, Illinois, Ohio, eastern Wisconsin, and north of Lake Superior. The smallest group, the Sioux, inhabited northwestern Ontario, Wisconsin, and Minnesota.

The Iroquois, who included the Huron, were farmers, but the amount of land they cultivated was limited. The Huron, for example, practised a slash and burn agriculture. The men cleared small plots of land using fire and stone axes to girdle the trees, and the women did the planting. Such agriculture was best done on light, sandy soils, which quickly became depleted of their nutrients and were susceptible to drought. Because of this the Indians had to move their villages every ten to twenty years, even though in some cases the villages were over ten acres in size and contained over two thousand people. Nonetheless, by the seventeenth century, the entire Huron population of thirty thousand people had no more than twenty-three thousand acres of land under cultivation.

Although agriculture was important to some of the native peoples in the Great Lakes region, they all relied heavily on fish and game for nourishment. Bear, beaver, and deer formed an important part of the food supply, as did the large numbers of wild birds, including geese and turkeys. The waters of the lakes, which contained trout, pike, sturgeon, and other species of fish were especially important for subsistence. Historian George Hunt, who has chronicled the life of some of the Great Lakes Indians, summarized the food sources available to the native peoples of Green Bay:

> Corn grew luxuriantly, and throughout the region the wild rice was abundant. There were deer at all times, and bear and beaver could be taken in winter, to say nothing of the wild fowl, which were so numerous that they could be caught in nets. And should the game fail, there were sturgeon in the lake and great schools of herring in the falls.[11]

The Chippewa Indians, who lived in what is now Michigan, also had a varied and plentiful food supply. In his study of Lake Huron history, Fred Landon said the streams running through the Michigan forests "teemed with fish, and waterfowl of many kinds were abundant during a considerable part of the year."[12] Clearings throughout the Michigan jack pine forest provided the Chippewa with an ample supply of both huckleberries and mosquito-free summer campgrounds.

The life of the native people demanded skill and ingenuity, but in the Great Lakes region they found an area of abundant natural wealth that sustained them well. Although they relied entirely on the land and the waters for sustenance, they had a small impact on the landscape. Their numbers were few, their tools relatively simple, and with the exception of some agricultural clearings they did not extensively modify the environment. They saw themselves as intimately connected to the natural world around them. They built homes and fashioned canoes suited to the terrain and natural conditions and to their nomadic or semi-nomadic lives. Describing the land ethic of native people, Vine Deloria Jr, a Sioux, wrote, "The land-use philosophy of Indians is so utterly simple it seems stupid to repeat it: man must live with other forms of life on the land and not destroy it."[13] When the first European explorers arrived in the Great Lakes region, they found a land that was not untouched by human hand but that, as one author noted, "bore more the imprint of nature than of man."[14]

But the character of the Great Lakes environment was soon to be altered. The arrival of Europeans in the sixteenth century quickened the pace of ecological change. The outsiders brought with them a concept of time different from the slow rhythms of nature that had guided previous changes in the region. They also brought a different concept of the relationships between humans and the environment – a concept that placed humans apart from the natural world. At first the impact of the Europeans was minor, but as exploitation and settlement accelerated the changes became more and more pronounced.

TWO

The Europeans Move In

WHEN HE SAILED from France in 1534 Jacques Cartier became one of a steady stream of European explorers who ventured west across the Atlantic in hopes of finding a passage to the riches of Asia.[1] An experienced navigator, Cartier left with orders from his king to look for gold, silk, and spices and to claim any newly discovered lands for France. Dreaming of fortune, Cartier instead found the mouth of the St. Lawrence River, the outlet of the Great Lakes and the entrance to a nineteen-hundred-kilometre waterway into the interior of North America.

On a second and larger expedition a year later, Cartier travelled six hundred kilometres up the St. Lawrence to the site of the present-day city of Montreal, where the narrowing of the river and the treacherous Lachine rapids blocked further progress. There he marvelled at the beauty of the region, and heard tales from the Indians of vast seas of unknown dimensions farther to the west. After staking his country's claim to the new lands, Cartier spent the winter camped along the shores of the St. Lawrence. In the spring of 1536 he returned to France with word of his "discovery."

Seventy years later, in 1607, Cartier's countryman, Samuel de Champlain, established a colony at Quebec. Only then did France begin to solidify its claim to ownership of the region. As one group of authors noted, after finding the entrance to the St. Lawrence and establishing a settlement at Quebec, "The French were handed the keys to the interior of North America."[2]

The energetic Champlain had first arrived in Canada in 1603 with a mandate from the King of France to chart the size and dimensions of the New World. In addition to his mapping, Champlain began a fur trade with the Indians and – as an act of goodwill towards the Algonquin and Huron people he wanted to trade with – went so far as to participate in raids against their traditional enemies, the Mohawk. The alliance forged between the French and the Algonquin and Huron proved profitable. Soon the luxurious furs of the beaver and other animals were on their way to Europe to feed the desires of wealthy Europeans. The exploitation of the resources of the Great Lakes environment had begun.

VOYAGEURS AND MISSIONARIES

In addition to his commercial interest in furs, Champlain was also a geographer and explorer. For a quarter-century he travelled the St. Lawrence River and surrounding area with Huron and Algonquin guides. He visited what is now upstate New York, Quebec, and parts of Ontario, and sent others off to explore other parts of the Great Lakes region.

In 1610 Champlain sent a tough and adventurous young man, Etienne Brulé, to live among the Algonquin tribes of the Ottawa River. He instructed Brulé to undertake explorations and learn the Indian ways. Although there is no written record of Brulé's travels, it seems certain that his Indian guides took him by canoe up the Ottawa River, across Lake Nipissing, and down the French River into Georgian Bay on Lake Huron. Thus the shores of Lake Huron, the middle in the chain of five Great Lakes, became the first of the lakes to record the footprints of Europeans.

Champlain, "the Father of New France," was the first to provide a written record of the lakes. In 1615 he retraced Brulé's route and, emerging from the mouth of the French River into the blue waters of Georgian Bay, was awed by the size of the body of water he found. In his journal he wrote, "Because of its great size, I named it Freshwater Sea."[3]

The riches of the fur trade, the continuing search for a passage to China, and the prospect of converting the Indians to Christianity motivated further French explorations of the region. Ironically, the three lakes farthest west – Huron, Superior, and Michigan – were explored before the more easterly lakes, Erie and Ontario. The Iroquois nation, hostile to the French and the Hurons, controlled the area surrounding lakes Erie and Ontario. To avoid Iroquois raiding parties, French voyageurs used the Ottawa River-Lake Nipissing-French River route to Lake Huron as the highway for travel and movement of furs.

These early explorers and missionaries faced a difficult life. Insects, disease, hunger, and fatigue were common. The letters of the missionary Father Joseph Le Caron, who accompanied Champlain on some of his journeys, provide a glimpse of the difficulties. "It would be hard to tell you," he wrote, "how tired I am with paddling all day, with all my strength, among the Indians; wading the rivers a hundred times and more, through the mud and over the sharp rocks that cut my feet; carrying canoe and luggage through the woods to avoid the rapids and frightful cataracts; and half starved all the while."[4] Despite these hardships, many willingly endured the difficulties in return for a role in the exploitation and exploration of the New Land.

Among those to explore the Great Lakes region was Jean Nicolet who, in 1638, paddled to Green Bay on the western shore of Lake Michigan. Believing he had found China, Nicolet stepped ashore among the doubtless surprised Winnebago Indians. For the occasion Nicolet had clothed himself in a brilliant Asian silk gown embroidered with patterns of birds and flowers. Disappointed that he had not found the Orient, he was, nonetheless, the first European to see the waters of Lake Michigan.

Thirty years later, Louis Jolliet and the black-robed Jesuit missionary Father Jacques Marquette returned to the Green Bay area and "discovered" a tributary of the Mississippi River just to the west of the bay. On a later trip to Lake Superior, Jolliet abandoned the traditional route of return to Montreal, down the French and Ottawa Rivers, in favour of paddling south on Lake Huron. He eventually found himself on Lake Erie, the last of the Great Lakes to be explored by the Europeans.

Prominent among the early explorers was René-Robert Cavelier, Sieur de La Salle, who built the first sailing vessel to ply the waters of the upper Great Lakes. In 1679 La Salle sailed west on Lake Erie in the forty-five-ton *Griffin,* built above the falls along the Niagara River. La Salle intended to establish a fur trading post in the western region of the lakes and to explore the strategic Mississippi River. Upon reaching Green Bay he decided to continue his explorations by canoe and filled the *Griffin* with furs and sent it back to Niagara. Father Louis Hennepin, accompanying La Salle, wrote in his journal, "They set sail on the 18th of September with a very light west wind, making their adieu by firing a single cannon."[5] The *Griffin* never did make it back to its home port and became, instead, the first of many Great Lakes maritime sailing mysteries.

The Great Lakes region provided abundant possibilities for traders interested in commercial exploitation of furs. But it was equally attractive to church leaders, who saw the potential to "civilize" – or convert to Christianity – the native inhabitants. With these explorers and missionaries came a new way of thinking about the Great Lakes environment. Unlike the native people, the European explorers did not view the region as a home. They saw it instead as a passageway to the East and a commodity or resource that could be exploited and turned to profit. This view of the land became the dominant force in shaping interactions between people and nature.

FUR TRADING AND EMPIRE BUILDING

While explorations of the Great Lakes progressed, the fur traders and voyageurs continued to send huge volumes of valuable beaver pelts across the Atlantic. As early as the mid-1620s the St. Lawrence River traders were handling twelve thousand to fifteen thousand beaver pelts a year.

The French, however, were not the only Europeans exploiting the fur resources of the region. The wealth also attracted the attention of the British and Dutch colonists on the eastern seaboard of the continent. Merchants from the Dutch colonies established trading links with the Mohawks, the Indian enemy of the French.

In exchange for furs the native people of the Great Lakes region received muskets, blankets, iron knives, and axes. They acquired something else as well: European disease. Beginning about 1634, small pox, measles, and influenza – diseases for which the Indian people had no defences – devastated the native population. The Hurons and Petuns were especially hard hit: in a scant five years illness reduced their population from a pre-epidemic level of somewhere between twenty and thirty thousand to only twelve thousand. A death rate approaching 50 per cent was also common among the other infected tribes.

Intertribal wars, intensified and exaggerated by the fur trade, further decimated the native population. The Iroquoian confederacy, composed of the Mohawk, Seneca, Cayuga, Oneida, and Onondaga tribes living south of Lake Ontario, began a series of raids against the Huron, Petun, Neutral and Nipissing, who lived north of lakes Erie and Ontario. As one historian coldly noted, "Coordinated planning and the effective use of muskets enabled the Iroquois confederacy to disperse the Huron tribes in 1647-49, the Petun in 1649-50, the Nipissing in 1649-51, and the Neutral in 1651-52." By the mid-point of the seventeenth century, the entire eastern portion of the Great Lakes had lost the majority of its original inhabitants.

The competition for furs among European merchants was intense. As a consequence, the French King Louis XIV decided that domination of the Great Lakes fur trade could only be achieved by establishing and expanding settlements and forts. So the French enlarged and added forts to some of their missions and trading posts, such as St. Ignace at Mackinac and St. Francis Xavier on Green Bay. They also built new settlements and forts in strategic locations, such as Frontenac (Kingston) in 1673 and Detroit in 1701. These trading centres not only formed the cornerstones of the empire but also marked the beginning of European settlement in the Great Lakes region. By 1750 the summer population at Fort Detroit had reached four hundred and five hundred more Europeans farmed in the surrounding area.

But competing interests sought control of the valuable region and it soon became the site of European battles. The British, colonial rivals of the French, eyed the spoils of the Great Lakes fur trade with envy. Beginning in 1755 they launched a series of attacks against French forts and outposts. After some initial military blunders, in 1758 the British attackers captured and burned Fort Frontenac on Lake Ontario. One year later the British successfully wrested control of the Great Lakes from the French when they overran the defences at Quebec City. Although most of the French settlers remained in the new land, the loss of Quebec marked the end of French domination in the region. Pioneers in its exploitation and settlement, the French left their imprint in the names of such cities as Sault Ste. Marie, St. Joseph, Detroit, and Marquette.

Under British control the fur trade continued to be the mainstay of the region's economy. The defeat of the French, however, did not end the conflict over regional power. The original inhabitants of the lakes, wary of the consequences of European settlement, fought increasingly numerous battles with the encroaching settlers.

On May 16, 1763, for example, the Wyandot, who lived in the area south of Lake Erie, captured and burned Fort Sandusky. Later in the same month the Potawatomi of southern Lake Michigan captured Fort St. Joseph. A few months later an ambush along the Port Road near Niagara Falls by the Iroquoian leader Seneca left 139 British dead. Despite these victories, inadequate munitions and weapons and smallpox epidemics turned the tide of war against the native alliance. The smallpox epidemic was induced by the distribution of clothing from the victims of the disease to native peoples and orders for this biological warfare came from the British commander. As a result, by the autumn of 1764 the Indians had capitulated and the spread of settlers, particularly south of the lakes, accelerated.

Warfare soon returned to the region as British and American troops fought the War of Independence. At the resolution of the conflict, "The Definitive Treaty of Peace" was signed in Paris in 1783, establishing the present boundary division between Canada and the United States. A dividing line ran down the middle of the lakes, the exception being Lake Michigan, which was granted entirely to the United States. An immediate result of the peace was a major wave of settlement in the Great Lakes region. Thousands of Loyalists fearing persecution at the hands of the victorious Americans moved north from the United States to settle along the shores of Lakes Ontario and Erie. In the eight years before 1800 the population of Upper Canada jumped from twenty thousand to sixty thousand people.

The peace proved less than definitive. The simmering conflict between Britain and the newly formed United States erupted again in 1812 and the Great Lakes held centre stage for a two-and-a-half-year war more notable for military blunders than brilliant strategy. In the course of the conflict the lakeside towns of Newark (Niagara-on-the-Lake), York (Toronto), Buffalo, and Port Dover were burned, along with buildings in the more distant U.S. capital of Washington. Warships also fought a bloody battle on Lake Erie. In the end the existing boundary divisions between Canada and the United States remained intact.

The resolution of the War of 1812 did not end conflict over control of the Great Lakes region. The westerly expanding population of the United States continued to fight numerous frontier battles with native people. The last of these conflicts, the Black Hawk Wars, was waged in Illinois in 1832. The defeat of Black Hawk and his warriors firmly established the dominance of the European traditions in shaping the future of the region. The original inhabitants of the basin, who by their subsistence lifestyle and limited numbers had left few scars upon the landscape, were in large part conquered. Shinguacouse, of Garden River near Sault Ste. Marie, expressed the views of his people in a letter to the governor at Montreal in 1849:

> When your white children first came into this country, they did not come shouting the war cry and seeking to wrest this land from us.... Time wore on and you have become a great people, whilst we have melted away like snow beneath an April sun; our strength is wasted, our countless warriors dead, our forests laid low, you have hunted us from every place as with a wand, you have swept away all our pleasant land, and like some giant foe you tell us 'willing or unwilling, you must go from amid these rocks and wastes, I want them now! I want them to make rich my white children, whilst you may shrink away to holes and caves like starving dogs to die.' Yes, Father, your white children have opened our very graves to tell the dead even they shall have no resting place.[6]

The conquest of the native people brought about a rapid influx of settlers and ushered in a new relationship between humans and the Great Lakes environment. The westward expanding settlers brought with them the dominant attitudes of Europe towards nature and life. They saw the land as a chaotic and uncivilized wilderness that needed humanizing and controlling. This belief became a driving force in shaping human activity in the region.

THE PRESSURES AND PAINS OF SETTLEMENT

The Lake Ontario basin was the first area in the region to be extensively settled by Europeans, and their initial task was the backbreaking work of clearing the forest for farming. One Canadian settler wrote home to his relatives, "The most that an immigrant can do the first year is to ... cut down the trees on as much ground as will be sufficient to plant ten or twelve bushels of potatoes, and to sow three or four bushels of grain."[7] Settlers were fortunate indeed to clear trees from two or three acres a year. Thomas Blaine, an unusually strong settler in Ohio, began clearing on March 18. By May 22, working as full a day as humanly possible, he had cleared what he judged to be four and a half acres.[8] An Ontario resident wrote in 1829, "In the course of ten years an industrious man may expect to have from twenty-five to thirty acres under improvement."[9]

To the European settlers, the forests were the enemy. Charles Twinning, an authority on the Great Lakes forests, wrote that to the settler the trees "stood against the progress of civilization and the establishment of smiling, prolific, sunlit fields."[10] The attitude had little to do with maliciousness but was an inherent part of the prevailing culture. "There was no question as to what was of ultimate importance – it was the land," stated Twinning. "And the trees were as much an obstacle in its utilization as was Indian title."[11] The settlers sold their marketable pine logs to the timber companies and simply piled and burned what they considered to be the undesirable trees, such as maple and beech.

The resulting fields were not neat and orderly. The settlers left tree roots in the ground to rot, and went about planting around them. The ground itself was often hard and untillable. In some instances farmers were forced to chop holes for seed with an axe. They planted crops by hand and cut grain with scythes. Chrysostom Ver Wyst, who settled land with his family near Little Chute, Michigan, recalled, "Our wrists would ache from digging and working in the hard, rooty ground."[12] At first the farmers' work generated limited rewards. It was years before the soil was properly conditioned and the tree roots fully rotted. Until then the harvested crops were often barely enough to feed the farming family.

Despite the hardships, the promise of land and the possibility of escape from the troubles of Europe prompted successive waves of immigrants to swell the Great Lakes population. The failure of potato crops in Ireland in 1818 and again in the 1830s and in 1845 triggered a flood of Irish immigrants. The collapse of a revolutionary movement in 1848 brought the "forty-eighters" from Germany. People from every European nation were represented in the droves of immigrants landing on the shores of the new land – an influx described by one author as "the most massive human migration in history."[13]

Some immigrants came from their place of origin as a group and their new settlements retained much of the culture of the old country. The names of many Great Lakes cities still bear the imprint of European heritage: places such as Holland, Berlin, Dunkirk, and Hamburg.

The work of all these settlers dramatically transformed the landscape. They cleared the forests, drained the marshes, and burned the prairies to make way for farms. Villages and towns developed, often around the site of a grist mill on the banks of a stream. The growing population built blacksmith's shops, stores, and sawmills, followed by taverns, hotels, schools, and churches. Whole communities quickly materialized from what only a few years before had been a mosaic of forest, wetlands, small meadows, and clearings. The region was alive with the buzz of activity.

Population increases made improvements in transportation necessary. The original European inhabitants, the voyageurs, had used a modified version of the Indian canoe to move furs and other goods. Capable of hauling five tons of freight and of being portaged by two men, the freight canoe was ideally suited to the natural water-highways of the Great Lakes region. But now entrepreneurs needed larger and more sophisticated methods of transport to move an expanding volume of goods. They also had to overcome natural barriers that impeded the movement of larger ships – the rocky rapids of the St. Lawrence River and the falls at Niagara, for instance. So energetic construction of canals around rapids, waterfalls, and land barriers began during the first few decades of the nineteenth century.

By 1825 the Lachine Canal was opened, allowing ships to bypass the rapids at Montreal. Workers completed the first Welland Canal around Niagara Falls four years later. In 1825 workers to the south finished construction of the Erie Canal, a dream of businessmen and politicians who wanted to connect the Atlantic port of New York with the Great Lakes. This 364-mile canal, running between Albany on the Hudson River and Buffalo on Lake Erie, helped ease the flow of goods and settlers between east and west, and the immediate effect was "explosive." The cost of moving freight from the lakes to the Atlantic dropped by 90 per cent and the boats came back from the coast filled with a human cargo. The Erie Canal meant that New York, instead of Montreal, became the outlet port for settlers' produce. Commerce flowed both east and west over the canal to and from the lakes, and each year the passage of goods and people increased in volume. The pace of human activity in the Great Lakes expanded rapidly.

The steamship era further improved the movement of goods and people westward. Walk-on-the-Water, the first steam vessel on the Great Lakes, began a regular run from Buffalo to Detroit in 1818. Immigrants arrived at Detroit by the thousands in the 1820s. "As both a way station and an embarkation point for the lands further west," wrote the Great Lakes historian Harlan Hatcher, Detroit "felt the tremendous stirrings of the continent."[14] By the 1830s, ninety ships a month, some of them with upwards of seven hundred passengers, were docking at Detroit.

At the conclusion of the Indian wars in 1832, settlement moved rapidly west, with four million acres of Michigan land sold off in 1836 alone. Farming, however, provided only one of the employment opportunities for the swelling population. The era of vast internal improvements – turnpikes, canals, and railroads – all required the labour of the immigrants, and the growing lumber industry could always use new recruits. There were abundant employment opportunities. Life could be hard, but work was plentiful for the venturesome and desperate people who had left Europe for the promise of the Great Lakes region.

"Little short of incredible," was how Milo Quaife, the author of a history of Lake Michigan, described the speed of settlement of the western Great Lakes states after 1830.[15] As late as 1820, Michigan, then including all of modern Wisconsin, had a white population of less than 9,000 people. By 1830 this had increased to 31,689, with most of the newcomers locating themselves in the lower Michigan peninsula. Just ten years later, reduced to its present boundaries, Michigan had a population of 212,000; by 1850 this had increased to 397,000; and by 1860 it had almost doubled, to 749,000.

The influx of settlers accelerated the cutting down of the forests, and the pioneers built new mills on available streams and established villages or expanded towns. Captain Basil Hall, travelling through the city of Rochester in the early nineteenth century, described the activity: "Hammers were clattering, axes ringing, machinery creaking."[16] In 1832 alone the city of Chicago increased from a population of one hundred and fifty to two thousand. By 1860 the population of Ontario had reached almost one and a half million.

THE SHAPE OF THINGS TO COME

The frantic settlement of the Great Lakes region was not without its non-human costs. The clearing of the land eliminated animals that required large tracts of undisturbed forests for survival. By 1850 the eastern subspecies of elk that was present when the French explorers arrived was extinct. The timber wolf, once common throughout the forests of the region, was reduced to only scattered groupings in Minnesota, northern Michigan, Wisconsin, and northern Ontario. As hunting, trapping, and habitat loss diminished the numbers of favoured fur species, the populations of wolverine, fisher, marten, and otter were all significantly reduced. Wild turkeys were hunted out of existence in large areas of the basin.

The beaver, once abundant throughout the Great Lakes region, became exceedingly scarce, which had a kind of multiplier effect on the land. As the authors of a report on the changes in wildlife populations in the region noted, "The drastic drop in the numbers of beavers must have initiated extensive environmental changes, for thriving beaver populations had created frequent forest clearings, elevated water tables, and created ponds that served as catchments for meltwater and eroding soils."[17]

One of the most tragic losses was the passenger pigeon, a once numerous bird found only in eastern North America. At one point its population exceeded three billion. As a boy, Potawatomi Indian chief Simon Pokagon watched huge flocks of passenger pigeons flying over his home in southern Michigan. Years later he wrote:

> I have stood for hours admiring the movements of these birds, I have seen them fly in unbroken lines from the horizon, one line succeeding another from morning until night, moving their unbroken columns like an army of trained soldiers pushing to the front.... At other times I have seen them move in one unbroken column for hours across the sky, like some great river, ever varying in hue; and as the mighty stream, sweeping on at 60 miles an hour.

But in the short space of twenty-five years the great flocks of passenger pigeons were driven to extinction by the advancing settlers. The destruction was straightforward. A participant in the killing describes the method:

> Our visit to the nesting grounds occurred as the young birds (squabs) were about to leave their nests. Scarcely a tree could be seen but contained from ten to fifty nests, according to its size and branches. We knocked the young fledglings out of their nests with long poles, their weak and untried wings failing to carry them beyond our reach, and gathering them up alive, put them into our wagon racks, made and fitted specially for the purpose. People came from far and near with every sort of conveyance to haul the young birds away and indulge their passion for killing upon the old birds, which also were slaughtered by the thousands.[18]

In addition to the growing population's visible and dramatic alterations to the landscape, the lakes, and the wildlife that the environment once supported, other subtle but important changes were taking place. The hydrologic cycle – the pattern of rainfall, the percolation of water into the soil, and evaporation – was slowly but noticeably being changed. By the 1880s farmers were complaining about summer drought and spring floods. They were finding that an increasingly large percentage of water did not seep into the soil but instead ran directly off their land. A sizeable organic layer is essential to ensure adequate percolation into the soil; the lumbering and farming of the European settlers, however, destroyed the humus, the dark and valuable soil formed from decayed leaves and other vegetative matter. Springs and creeks dried up entirely during the summer and rivers became sluggish. In the spring the surface run-off from rain and melting snow on treeless ground was rapid and brought flooding to record levels.

The accelerated runoff increased soil erosion, which was particularly damaging along the lake fronts and on the steeper, more recently cleared lands. Once-clear rivers and streams turned muddy and brown. Masses of silt covered up the gravel on river and lake bottoms, eliminating the traditional spawning sites for fish such as sturgeon, muskellunge, and lake whitefish. Silt also covered over the aquatic vegetation rooted in shallow bays, reducing the quality of these spawning areas for pike and sturgeon.

On large fields of sandy soil where white pines had once grown there were now huge sand drifts and blowouts. Desert-like wastelands developed in areas such as Norfolk County north of Lake Erie and Simcoe County east of Georgian Bay, where, only a few years earlier, magnificent pine forests had grown. On some farms in southern Ontario drifts from blowing sands covered over thirty hectares.

To power the grist mills that were needed to grind wheat and other grains, the settlers constructed dams along the thousands of streams and rivers flowing into the Great Lakes. The dams in turn changed the character of the water flowing to the lakes. Direct sunlight on the impounded water increased its temperature,and the dams blocked the migration of river-spawning fish. Imperceptibly at first, but with a gradually increasing force, the Europeans and their various enterprises forever changed the character of life in the Great Lakes region.

THREE

Trees, Fish, and Rock:

The Resource Bonanza

RAPIDLY DEVELOPING industrial economies require building materials, and the Great Lakes region was blessed with a building resource of seemingly inexhaustible proportions: its forests. Some trees towered over two hundred feet, with diameters of six feet, making building material both abundant and accessible throughout the basin. Before the settlers could clear the land for agriculture, lumber workers often entered an area to cut down the available timber. Beginning with the area around lakes Erie and Ontario, a rapid exploitation of the forests took place.

Early in the nineteenth century a frantic rush for all available supplies of standing timber accompanied the fantastic growth of the forest industry. The blitz quickly pared down the forests north of Lake Ontario and Lake Erie so that by 1861 forest covered only 77 per cent of the land directly north of Lake Erie. Within another sixty years lumber demand had reduced the forest cover to a meagre 19 per cent of the land area. By 1897 one observer could describe the western New York drainage area of Lake Ontario as "almost totally deforested."[1]

Ports such as Oswego, New York, on the south side of Lake Ontario, became important centres of lumber activity. In 1840, 1.9 million feet of Canadian lumber were off-loaded at Oswego. Less than ten years later, in 1849, 44 million feet entered the port. Soon after the figure increased to 60 million feet.

The settlers' demand for lumber products was exorbitant; each year the growing North American populations used 350 cubic feet of lumber for each man, woman, and child. The lumber industry also exported substantial amounts of timber, especially from Canada to Britain. "The speed with which the vast regions of the Great Lakes country were stripped in huge lumber commerce is remarkable in world history," wrote William Ellis, the author of a history of the Great Lakes region.[2] The lakes themselves provided an effective means of transporting the lumber from forests to market. In 1840 fifty vessels worked the lumber trade. By 1885 there were some seven hundred clippers and steamers.

As the eastern forests became denuded, the lumber industry shifted its attention farther west, stripping the forest cover from the watersheds of each of the Great Lakes in succession. At the peak of lumbering activity, the shores of Lake Huron were piled high with a carpet of sawdust and shavings. The Shiawassee, Cass, Flint, and Tittobawassee rivers each sent a torrent of timber down to Saginaw Bay. At Wiarton, on the south end of Georgian Bay, twenty-five rail cars a day were loaded with lumber from the surrounding lands. The harbours and waterfronts of the lumbering towns were vast seas of floating logs.

After the most accessible stands of white pine in New York and Pennsylvania had been cut, Michigan began supplying parts of the eastern market by way of Lake Erie and the Erie Canal. About nine million acres of Michigan forest were composed of stands of valuable white pine. As W.L. Clements wrote in *Michigan History Magazine*, "During these years was begun and carried to consummation, without regard for conservation or the rights of future generations, the destruction of Michigan's white pine forests."[3]

Driven by the expansion of settlement into the prairie states, the lumbering industry spread into Wisconsin and Minnesota. Lumbering began in this area around 1850 and after the Civil War the cutting increased and quickly became a major industry. One historian described the lumbering at the conclusion of the war as a "mass attack on the forests of the upper Great Lakes states."[4] As cutting outstripped local demand, the lumber was increasingly exported west by ship and rail to the expanding population of the mid-western states. A.H. Lawrie and J. Rahrer, biologists who compiled a history of human impact on Lake Superior, described the deforestation of the Lake Superior watershed:

> As long as they lasted, white and red pine were selectively cut and the volume of timber taken from the area was staggering ... The exploitation was ruthless and complete. By 1896 pine was becoming scarce along most of the south shore.[5]

A single company alone shipped seventy-five million feet of Lake Superior lumber and thirty million shingles in 1870. Some twenty-five years later, the annual production from two counties on the south shore of Lake Superior was 154 million feet of pine lumber and six million shingles.

The forests seemed inexhaustible. In 1852 Wisconsin congressman Ben Eastman reported to the U.S. House of Representatives that in the Great Lakes region, "There are interminable forests of pine, sufficient to supply all the wants of citizens ... for all time to come."[6] The vastness of the forests meant that the lumber industry gave little thought to forest management or conservation.

Forest fires were a devastating consequence of the lumber industry's indifference to conservation. Although fires were always an integral part of the forest ecology, their size and frequency increased dramatically after European exploitation began. Early pine logging used only 20 to 30 per cent of the wood in the tree and workers left behind tops, limbs, and other wood waste on the forest floor. The wood wastes provided kindling for frequent fires, the understory was burned, the forest litter destroyed, and the soil previously held in place by plant roots easily eroded. The regeneration of the forest in such burned-over areas was extremely slow.

In 1871 one such fire burned out of control through northeastern Wisconsin. A prolonged drought through the summer and early fall had made the western Great Lakes forests ripe for the ravages of fire. The burning of brush and leftover timber by farmers, railway crews, and loggers dangerously tempted fate. Throughout the autumn, numerous small fires had started, but on the night of October 8 southwest winds fanned the flames into a raging inferno. The carnage that followed reached unbelievable proportions. The fire moved rapidly up the eastern and western shore of Green Bay and scorched entire towns. One eyewitness described the scene as "a hurricane, seemingly composed of wind and fire."[7] The fire reduced the town of Peshtigo to cinders, and people and animals unable to outrun the flames simply burned in their tracks. A monument to the eight hundred lives lost in "the tornado of fire" now stands in the Peshtigo cemetery.

The wrath of the fire continued to be felt long after the flames from the blaze had subsided. Reverend Peter Pernin, who survived the fire by submerging himself in a creek, told about "whole forests of huge maples, deeply and strongly rooted in the soil" – the trees "torn up, twisted and broken ... and their branches reduced to cinders and their trunks calcined and blackened." Root systems completely collapsed, leaving the earth vulnerable to erosion. Formerly clear, cool streams and rivers were darkened by the onslaught of eroded soil washing easily from the treeless land. Fish were smothered and died in "immense numbers."[8]

Although not as immediately destructive to the landscape and the lakes as fire, other forestry practices also damaged the Great Lakes environment. For centuries the abundant white pine had produced bumper seed crops at intervals of three to ten years. These seeds regenerated new white pine forests following periodic fires or blowdowns. However, the heavy cutting of the pines reduced the available seed supply. There was simply not enough seed to regenerate the pine forests and, as a consequence, aspen, birch, and scrub species replaced the pines.

Aspen in particular proliferated in the cut and burned lands. Previously only a minor component of the Great Lakes forests, aspen regenerates via underground suckers, giving it a competitive advantage in the cut and burned areas. The regenerating forests became vastly different from the forests that had evolved in the region over thousands of years.

The methods of lumbering used also had destructive ecological effects, particularly on fish. Workers transported timber from forest to sawmill by "driving" the rivers. They did most of the cutting in winter when the snow could be packed smooth, allowing the logs to be more easily pulled from the forest by horses. In the spring, when the melting snow swelled the rivers, they drove the accumulated logs to the mills using the fast-flowing current. River hogs, loggers armed with peaveys, followed the tossing, tumbling logs on their journey down the river, unlocking any jams. The driving of rivers was an efficient means of moving timber from the forest, but caused extensive scouring of stream beds, killing fish eggs and damaging the spawning habitat.

The harm did not end with the driving of rivers. Once logs had completed their journey to the lake, loggers formed them into rafts and towed them to the sawmills, in some cases a trip of over one hundred miles. One exceptionally large raft, composed of 91,700 logs, crossed Lake Huron from the mouth of the French River to Saginaw, a distance of 230 miles. On these journeys the bark and wood chips that flaked off the logs would litter the lake bottom with coarse woody material, further damaging fish spawning habitat. Some river and lake bottoms became virtual biological deserts.

The sawmills themselves generated huge mountains of sawdust that severely polluted the inshore areas around the mills. By 1880 observers noted a 90 per cent decrease in whitefish spawning on Green Bay, and attributed this to the huge volumes of sawdust coating the bottom of the Oconto River and an area extending two miles into the bay. Sawdust pollution also severely reduced whitefish spawning in Saginaw Bay on Lake Huron.

The valuable timber reserves of the region were an essential raw material in the development of the Great Lakes' cities and towns. But such rapid exploitation of a slowly regenerating resource could not last forever; the lumber boom was brief. By the twentieth century almost all the virgin timber had been cut. In 1944 author Milo Quaife described the landscape at the close of the lumber boom:

> The finest white pine and hardwood forest in the world is now a man-made desert of fire blasted stumps and slashings. The Muskegon River, once choked with timber as far as the eye could see, is empty.... There are rotting piles and moss-covered wharves where once echoed the busy refrain of 47 giant sawmills, adding night and day to the vast board pile surrounding Muskegon Lake.[9]

As it used up the supplies of timber, the lumber industry continually moved elsewhere until there were simply no more accessible reserves of wood left. Today, broken-down buildings and huge piles of sawdust are all that is left at the site of many of these former mills.

THE RISE – AND RESULTS – OF COMMERCIAL FISHING

Like the forests, the plentiful fisheries of the Great Lakes astounded early visitors. *The Relations,* a journal of the experiences of the Jesuit missionaries published each year in France, noted that on the south shore of Lake Superior, "A single fisherman will catch in one night twenty large sturgeon, or a hundred and fifty whitefish, or eight hundred herring in one net."[10] The rapids at St. Marys, where Lake Superior empties into Lake Huron, was the annual site of a large native fishing camp that provided the Indians with a full supply of food. The Jesuits at Sault Ste. Marie wrote of whitefish running so thick in the St. Marys River that a man standing in the water could reach out and simply grab a thousand or so. The earliest descriptions of the Great Lakes reported major spawning runs of trout and salmon and other kinds of fish in estuaries and tributaries.

Over one hundred and fifty species of fish carved out an ecological niche in the waters of the Great Lakes. The central waters of the open lakes, primarily deep, cool, and oligotrophic, were home to lake trout, sculpin, and lake herring. Whitefish, an important component of the diet of many Indians, moved between the open lake and their near-shore spawning areas. The Atlantic salmon, the only salmon native to the Great Lakes, also moved from its wanderings in the medium depth and deep waters of Lake Ontario to near-shore areas before it swam up suitable spawning streams. In the shallow, biologically-productive parts of the lakes, such as the western end of Lake Erie, the Bay of Quinte on Lake Ontario, Saginaw Bay on Lake Huron, and Green Bay on Lake Michigan, species such as pike, yellow perch, walleye, and bass thrived.

At first, settlers of Lake Erie, like the Indians before them, used hooks, simple seines, and spears to capture whitefish, walleye, and other fish. During the early nineteenth century, the scale of the fishery increased as the abundant and seemingly inexhaustible nature of the resource led to the establishment of commercial fishery operations.

Commercial fishers on Lake Ontario targeted Atlantic salmon, lake trout, whitefish, and walleye beginning around 1830. J. Pickering, living on the shores of Lake Ontario, wrote in 1831, "Two persons in a canoe with a spear and torch will sometimes kill eight or ten barrels of salmon in one night" – with each barrel holding about 200 pounds.[11] By mid-century a substantial commercial fishery had also developed on Lake Erie, most intensely in the southern, U.S. half of the lake.

On Lake Huron, Captain Alexander MacGregor of Goderich carried on one of the earliest and most successful operations. In the shoals off the Fighting Islands, situated near the Saugeen Peninsula, MacGregor discovered an abundant source of whitefish and herring in 1831. MacGregor's practice was to station a person to watch out from a high tree near the shore, with a clear view of the lake. When a "bright silvery cloud moving through the water" came into view the lookout would yell to the fishing party on shore and they would set out with their boats and nets to encircle the fish.[12] Historian Norman Robertston, writing at the turn of the century, reported, "When the fish commenced to feel pressure from the narrowing of the net" as they were hauled to shore, "the scene was one long to be remembered. There in a small area were trapped thousands and thousands of fish, sufficient possibly to fill five hundred or a thousand barrels." When the fish were manoeuvred to shore they were thrown out onto the beach by a man standing in the middle of the net who "imprisoned fish, scoop in hand" and transferred the catch "from their native element to land where they formed a splendent mass, flapping and gasping life away. At times the catch was so large that the landing of the fish was extended over three days, so that none be lost through the inability of the curers to handle so many."[13]

The first commercial fishery operation on Lake Michigan started as early as 1843. Whitefish was the principal target and the plentiful fish allowed a rapid growth of the industry.

Hudson Bay trading post personnel were the first to commercially exploit the fish stocks of Lake Superior. The journal of the Michipicoten trading post on the north shore recorded that local fishers caught thirty-six barrels of lake trout in the spawning run of September 1829.[14] Other fishing operations developed with the arrival of the first settlers along the south shore around 1850. The initial catch of fish from Lake Superior was primarily used to meet the needs of the expanding population associated with mining and lumbering. However, historian Grace Lee Nute reported: "As time passed, an increasing fraction of the catch was exported to the booming U.S. midwestern cities to the south, at first by water then, as rail service in the area developed, by both rail and water."[15]

The techniques used for catching fish depended on the lake and the species being sought. Seines and pound nets were among the earliest fishing techniques. During the mid-years of the century, gill nets gained in popularity among the fishers. Still used today, gill nets consist of a curtain of mesh stretched out in the lake. The fish attempts to swim through the net and when it realizes the opening is too small to go through it reverses and catches its gill covers or fins on the mesh.

In the second half of the century the commercial fisheries on the Great Lakes underwent progressive modernization of equipment and practices. Manufacture of improved linen and cotton nets began in the early 1870s. In 1871 the first steam-driven vessel appeared on the lake and by 1890 some of the boats were using steam power to lift their nets. The increased ease of setting and lifting nets allowed more work to be done by each fisher.

But as fishing intensity increased, and human-initiated changes to the environment accelerated, various fish species began to suffer. On Lake Erie, the most productive and most heavily fished of the Great Lakes, sturgeon numbers dropped to record low levels during the 1890s – largely due to purposeful overfishing. The sturgeon was large, sometimes weighing over eighty kilograms, and its external body armour easily tore nets set for smaller fish. To rid their fishing grounds of the sturgeon the fishers devised heavier nets. When they caught the sturgeon, fishers often "piled them like cordwood on the beaches, dousing them with oil, and burning them."[16] Because the long-lived sturgeon (some live 150 years) does not reach sexual maturity until somewhere between fifteen to twenty-five years old, its numbers were rapidly reduced.

Other fish species also declined. Black-finned and short-nosed ciscoes, herring-like fish that inhabited the lakes' deeper waters, were heavily fished; by 1900 the large black-finned ciscoe was commercially extinct on Lake Ontario. Important whitefish spawning shoals on lakes Huron, Michigan, and Superior became depleted as well, forcing fishers to move to other areas. Some walleye populations also declined. In 1871 fishers in Lake Huron's Saginaw Bay had seined walleye at a rate of several tons per haul. A few years later conservation officers reported only small runs of this fish. In 1871 the decline of fish on Lake Michigan prompted the Wisconsin Commissioner of Fish to conclude, "It is evident to all that the waters of Lake Michigan are being gradually depleted of fish."[17]

Despite the loss of some fish species, the Great Lakes fishery maintained its high level of production. Increased fishing effort, increased efficiency of gear (for example, smaller meshes, finer twine) and a shift to catching immature fish and seeking new fishing grounds all contributed to maintaining the total level of fish landings. By the turn of the century ten thousand people were involved in commercial fishing in the Great Lakes, double the number twenty years earlier. Commercial fishing was so intense that one observer remarked, "Considering the immense quantity of netting employed in so small an area as Lake Erie, it is surprising that any fish are left."[18] The composition of the catch changed, and how long the remaining stocks would last became a serious question.

The Atlantic salmon was one of the first major fish species to disappear. A large predator once plentiful in Lake Ontario, the Atlantic salmon spawned in tributary rivers and streams and after two years' growth moved downstream to the open waters of the lake. During their fall spawning runs, salmon were so abundant in Wilmot Creek that "men killed them with clubs and pitchforks" and "women seined them with flannel petticoats."[19] Restricted to Lake Ontario by the upstream barrier of Niagara Falls, the Atlantic salmon had been an important food for the Indians and early settlers. But by 1896 it was extinct in the Great Lakes.

The decline of Lake Ontario's Atlantic salmon occurred in a matter of decades. We know that salmon were abundant in the lake into the 1830s, but by 1866 their population had dwindled so much that a biologist attempting to capture salmon for a program of artificial propagation had difficulty obtaining enough parent fish for his operations. Hatchery releases led to a limited recovery in the 1870s. But the last salmon seen in the lakes was "a pair of 7 or 8 lb. fish" in Wilmot Creek in 1896.[20] No one single factor brought about their disappearance: like many other environmental disasters, the elimination of a species is often caused by a combination of stresses. The pressures of new and intense commercial fishing undoubtedly played a part in the demise of the salmon, but there is evidence that other activities of the early settlers unwittingly played a part as well.

After clearing land, one of the first things the settlers did was construct grist mills and saw mills. For instance, settlers along Toronto's Humber River and its tributaries had established thirteen mills by 1818 and ninety by 1860. To power the mills they built dams. An effective energy source, these dams nonetheless blocked the upriver movement of spawning salmon. The clearing of land had also increased soil erosion, so that the salmon's gravel spawning beds became covered with silt. The clearing also led to lower stream levels in the autumn, which further reduced the spawning areas and created winter anchor ice. When the ice moved during the spring breakup, it destroyed incubating eggs. The elimination of the forest cover from the banks of many of the spawning rivers reduced areas of shade where the salmon could lurk and increased the summer temperature of the river. Incubating salmon eggs have very sensitive temperature requirements, and increases of only a few degrees proved disastrous. The combined effect of all these factors left the Great Lakes poorer by one important species.

Throughout the Great Lakes other fish species suffered similarly from the damaging effects of land clearing. Lake whitefish had successfully spawned in the Detroit River and Maumee Bay of the western basin of Lake Erie until about 1890. By 1890, pollution had put a stop to the spawning runs in the Detroit River. The once clear waters of the Maumee River were turned a murky brown by the increasing load of silt eroding from deforested land. The whitefish of Green Bay also suffered a dramatic decline as a result of sawdust pollution from the lumber industry that surrounded its spawning habitat. Only the lake-spawning populations were left to support the expanding fishery.

During the late nineteenth century other changes took place in the Great Lakes fisheries. Fish species never before found in the lakes showed up in the catches of commercial fishers. Some of the new species were intentionally introduced to the lakes to compensate for the diminished quantities of native fish. Other non-native fish gained entry to the lakes by using the canals that now circumvented natural barriers to movement through the chain of lakes. One of these, the alewife, came via the Erie Canal and was abundant in Lake Ontario by the 1870s.

Rainbow trout were stocked in Lake Huron and Lake Michigan as early as 1880, and brown trout were planted in Lake Michigan in 1883. Carp, considered a delicacy, was another of the non-native fish introduced into the lakes. Attempts to establish both Pacific and Atlantic salmon in Lake Michigan were made between 1872 and 1932, but without success. During this period the fishing industry also started to augment native fish stocks with fish produced from hatchery operations.

By the end of the nineteenth century all of these factors – overfishing, disruption of habitat, pollution, and intentional and unintentional introduction of species – had brought drastic change to the fisheries of the Great Lakes. The bountiful fish resources that had once so effectively supplied food for the native Indians had been dramatically altered – largely for the worse.

ORE FOR THE TAKING

Mineral resources – vast, rich, and varied – are found throughout the Great Lakes region, but in especially large quantities in the Sudbury basin of Ontario and on the shores of Lake Superior. Copper, silver, gold, and, most significantly, iron ore are the major minerals of the Lake Superior area. The rocks of the Sudbury region north of Lake Huron contain nickel and copper, together with the precious metals platinum and gold. Discovered during construction of the transcontinental Canadian Pacific Railway in the 1880s, the Sudbury nickel deposits are among the richest in the world.

The early explorers were aware of the mineral wealth of the Great Lakes region. Etienne Brulé returned from some of his travels with copper nuggets. Father Claude Allouez, who established a mission on Lake Superior, wrote in the 1668 volume of the Jesuit *Relations*, "One often finds at the bottom of the water pieces of pure copper, of ten and twenty livres weight."[21] Other explorers also collected large chunks of copper from the Lake Superior region. During the mid-nineteenth century the geological search for commercially exploitable minerals increased. Singlemindedly focused on their quest for minerals, geologists sometimes burned down forests to expose the rocks. In 1848 Peau de Chat, chief of the Fort William Indians, complained to an Indian affairs official at Sault Ste. Marie, "The miners burn the land and drive away the animals, destroying the land."[22] Geologists saw everything that stood between them and the valuable rock they sought – the trees, soil, wildlife – as a distinct obstacle.

Miners began work on the copper deposits of Lake Superior in 1845 and produced 26,880 pounds that year. The Canadian Copper Company began operations at Sudbury in 1886. The early copper and nickel mining and smelting operations were neither clean nor efficient. Once mined, the copper and nickel ores were dumped onto layers of burning cordwood where they smouldered for months. This was a cheap method of reducing the ores, of smelting, but some of the copper, nickel, and other metals nonetheless washed away during rains.

The inefficiency of the system was, however, not its greatest shortcoming. In the Sudbury basin, for example, the minerals are locked in rocks containing large amounts of sulphur. The smelting process released this sulphur as sulphur dioxide gas, which formed large putrid clouds that drifted heavily over the surrounding land

and had a quick and devastating effect on vegetation. An early traveller to the region described the results: "A more desolate scene can hardly be imagined than the fine white clays or silt of the flats, through which protrude, at intervals, rough rocky hills, with no trees or even blade of grass to break the monotony."[23] Once again a chain reaction was set in motion. When the fumigated vegetation died, vulnerable soils became easily eroded. A forty-square-mile area around Sudbury became a wasteland of blackened rock where even soil micro-organisms, needed to decompose the twisted stumps of dead trees, could not survive. Many of these stumps remain in the area today, a ghostly reminder of earlier mining activity.

While copper was the first mineral to be extracted from the Great Lakes region, iron ore became the most prolific and important. The discovery and exploitation of iron dramatically influenced the pattern of development in the region. The chief deposits of iron are south and west of Lake Superior. The most productive of the iron mining districts is the Mesabi region in Minnesota, which by 1890 was producing more than all the other iron ranges combined. There were smaller deposits at the eastern end of the lake near Wawa, Ontario. Geologist Douglas Houghton discovered iron in southern Lake Superior around 1840. Houghton believed that the hematite (iron-containing rock) he found in Michigan along the shore of Lake Superior was not rich enough to justify mining, but a few years later a party of surveyors guided by Indians followed an unusual deflection of their compass to large outcroppings of ore further inland from Houghton's discovery. They organized a company and began mining at the site, located near Teal Lake in Marquette County, Michigan. They made their first shipment of hematite from the mine in 1852.

To smelt the ore, many of the iron mines used charcoal made from the burning of hardwood trees. At one time various mines in the upper peninsula of Michigan operated over thirty iron furnaces and hundreds of charcoal kilns. This use of hardwood forests for charcoal placed further strain on a resource that was already being severely depleted by the lumbering operations.

For some of the iron mines it proved to be better economics to use coal for smelting the ores, which meant that the companies shipped the raw mineral south along the lakes to the Ohio Valley. The seven-metre drop from Lake Superior into the St. Marys River at the Sault, however, posed a formidable barrier, so plans were laid for the construction of the Sault Ste. Marie locks. In his inaugural address to

Congress in 1851, U.S. President Millard Fillmore observed, "A ship canal around the falls of St. Marys of less than a mile in length, though local in its construction, would be national in its purpose and benefits."[24] A scant four years later workers finished building the canal, and the first of many shipments of Lake Superior ore passed through the Sault on its journey to the steel mills of Cleveland.

The construction of the Sault locks completed the water highway from the head of Superior to the Atlantic. Heavy ore shipments could now exit easily from Lake Superior to the iron smelters on the lower lakes. Yet there was more to come. The demand for iron during the U.S. Civil War stimulated mining even more, and brought ore production in the Superior District to a scale unparalleled in history. During the next century the vast deposits of ore would fuel the incredible industrial and capitalist development already taking place around the Great Lakes – a development with its own important implications for the health of the regional ecosystem. Human endeavours in the pursuit of wealth and a particular type of progress were opening up yet another chapter in the history of Great Lakes change.

FOUR

Power, Pollution, and the Pace of Change

NIAGARA FALLS: on both sides of the border, the honeymoon capital of the world, home to numerous game and amusement parks and assorted tourist traps. The main attraction: watching the powerful current of the Niagara River pour over the falls and crash onto the rocks below. The experience is humbling, a seemingly eternal and dramatically powerful reminder of the beauty and strength of the forces of nature.

A few hundred yards away from the falls at Niagara is a different yet no less powerful reminder, this time of the human changes that have taken place in the Great Lakes region during the twentieth century. Moulded into the rocks of the Niagara gorge not far from the mist-filled spectacle of the falls are the giant concrete spillways of electrical generating stations. These spillways are the creation of engineers and entrepreneurs who saw something other than the beauty of nature and their own humbleness in the falling water of Niagara Falls. They saw instead an opportunity that could be turned to profit and political power.

The invention of the dynamo, which could convert the flow of falling water into electricity, provided the means of seizing this opportunity. By 1895 engineers had designed electrical generating plants to harness the huge electricity potential of the falls. Electricity soon began to flow from giant turbines and by 1896 the city of

Niagara Falls, New York, was using over 13,500 horsepower of electricity. Industries in the area consumed half the electrical output from the Niagara power project, using previously unknown electrochemical processes. An aluminum company, a carborundum company, the Dunlop Tire and Rubber Corporation, three oil companies (Gulf, Richfield, and Frontier), and assorted chemical companies, including E.I. Du Pont de Nemours, built plants along the Niagara River close to the source of power.

Guiding the actions of the twentieth-century promoters of industrial growth was a belief in the human ability to manipulate and control nature. To describe their activities the new planners and managers used words like "taming" and "harnessing." In a speech to the Royal Society of Canada at the turn of the century, a prominent Canadian engineer described water power as "white coal." Here was an even cheaper and better energy source that would further stimulate industrial development in the region.

At the end of the nineteenth century it was evident that the coming century would be an era of immense economic growth in the Great Lakes region. Industrial activity boomed, fuelled by easily available energy, rich minerals, forests, agriculture, and fish. There was in place a transportation system, based on the lakes themselves, which was second to none in its ability to move vast quantities of goods quickly and inexpensively. Towns and villages such as Chicago, Cleveland, Sault Ste. Marie, Detroit, and Toronto had grown into major cities, and for much of the new century life in this "heartland" would seem intensely aglow with the new prospects and benefits of an advancing industrial society.[1]

THE GROWTH AND GRASP OF INDUSTRY

Just as the beaver had been the driving force behind an earlier economic incursion into the region, so too iron ore, fashioned into steel, became the driving force of the Great Lakes economy of the twentieth century.

The rise of the city of Gary, Indiana, near Chicago, illustrates the quick impact of steel on the Great Lakes economy. In the first decade of the century, a newly-constructed U.S. Steel plant hired on workers by the droves, but they needed homes. To satisfy that need, in 1906 the town of Gary went up over sand dunes on the southern edge of Lake Michigan and quickly developed into one of the region's largest cities. Other industrial development sprang up around the

Gary steel mills. Within only a few years, according to historian Milo Quaife, vast industrial development had "pushed steadily eastward around the shore," leading to "congeries of cities whose roaring furnaces and bellowing smokestacks redden the sky by night and blacken it by day."[2] The pace of industrial growth was relentless. Dunes, swamps, forests, and rivers – anything that stood in the way of the advancement of industry – were bought out, bulldozed, and buried.

The region's rapid industrial expansion demanded larger ships, better loading and docking facilities, and improved locks and channels. The Sault Canal, originally built with two locks, each three hundred and fifty feet long, seventy feet wide, and nine feet deep, had to be expanded and improved. Work began in the 1870s and was finished in 1896 but the advent of even larger ships soon necessitated further improvements. Workers began construction of a new canal and two new locks at the Sault in 1907 and 1911. When they were completed in 1914 and 1919 the expanded locks were 1,350 feet long, eighty feet wide and twenty-four and a half feet deep. The improved transport system was a success: in 1870, 830,930 gross tons of ore went through the Sault; in 1900 tonnage had increased to 19,059,393; and in 1920 it had reached 58,527,226.

While iron was the life blood of the Great Lakes economy at the turn of the century, it was by no means the sole focus of industrial activity. To supply the needs of an expanding population, manufacturing was taking on greater and greater significance. Detroit factories built steam engines, buggies, and wagons. Collingwood on Georgian Bay was a centre of shipbuilding activity, and cities such as Buffalo, Chicago, and Toronto produced a whole host of consumer commodities. Pulp production for paper manufacture overtook the waning lumber industry as the major user of the forests.

Meanwhile, the chemical industry, whose activities would later have ominous implications for the health of the Great Lakes, established itself in Michigan. In August 1890, twenty-four-year-old Herbert Dow, a Cleveland chemical engineer, reportedly emerged from a railcar onto the train platform in Midland, Michigan, with one hundred dollars in his pocket and a head full of plans for starting a chemical plant. Producing chlorine and bromine from local salt wells, Dow eventually moulded his company into one of the world's largest chemical producers.

The chemical industry cast its twentieth-century tentacles out in all directions, and one of its lasting links was made with the agricultural sector, which continued to play an important part in the economic growth of the region. The prevailing ethic that people could control and manipulate nature fostered a rapid increase in the use of chemicals for farming, establishing a dependency with implications that would not be felt until much later. In the short term the increasing mechanization and application of science and technology to farming improved crop yields, feeding a growing population.

Farmers also managed to increase agricultural production by draining off wetlands. The Great Black Swamp, for example, a 1500-square-mile wetland in northwestern Ohio, was crisscrossed with drainage ditches and tile drains by the end of the nineteenth century. In the words of one observer, the area was transformed from "a useless, obstructive morass into one of the most productive regions in Ohio and the corn belt."[3] Highways through the area today pass by mile after mile of corn field: a monotonous testament to the magnitude of the landscape transformation.

With the growth of industry and agriculture came more people. By 1900 the Lake Michigan basin alone supported over four million people. At the same time the Lake Erie watershed was home to over three million people. With their belief in the ability and right to control and modify nature to human ends, the future looked bright. The human mastery of science and technology appeared to be growing yearly and there seemed few limits to human ingenuity.

The growing wealth and good fortune of the area were glowingly expressed in a 1918 report of the International Joint Commission, a binational commission established to explore issues concerning the boundary waters between Canada and the United States:

> The people of both countries possess, in the splendid immensity of the series of waterways through which so much of their common boundary passes, a heritage of inestimable value. Millions of people dwell in their watersheds. Along the banks of the rivers and Great Lakes communities which a few years ago were mere villages are now in population, in social and industrial development, among the most important on the continent.[4]

But beneath the optimism of the new century there were signs that the environment on which this development depended was suffering. Individuals and groups began to warn of the deteriorating condition of the lakes and the basin landscape. Of particular concern was pollution by domestic sewage, which was for the most part discharged directly into the lakes and rivers. In the early years of the nineteenth century a Toronto newspaper had reported that "all the filth of the town – dead horses, cats, manure, etc." was "heaped up together on the ice, to drop down in a few days to the water."[5] The rivers flowing through the cities of Detroit, Toledo, and Toronto were, by the end of the nineteenth century, "open sewers" carrying the wastes of town to the lake.

THE IJC AND THE BATTLE AGAINST SEWAGE

In 1918 the International Joint Commission delivered a scathing attack on the quality of the Great Lakes waters. According to the commission the shore waters were "unsightly, malodorous, and absolutely unfit for domestic purposes."[6] Raw sewage carried death and disease in the form of cholera and typhoid bacteria. Drinking water drawn from the lakes was badly contaminated. In the first decade of the twentieth century, 525 persons out of every hundred thousand in Duluth, Minnesota, became ill with typhoid fever, and 52 died. In Ontario, an incredible 446 deaths per hundred thousand people were caused by typhoid fever. Between 1890 and 1905 an average of 85 deaths a year were attributable to typhoid in the Great Lakes region inclusive of Rainy River, west of Lake Superior.[7]

Robert Kedzie, a Detroit doctor and professor of chemistry, was one of the first to raise the alarm about the spread of disease by the contamination of water supplies with human sewage. "When the population along the river above Detroit becomes greatly increased the waters of the Detroit River will become unfit for domestic and potable use," Kedzie wrote in 1898.[8] A Toronto paper, *The News*, issued a similar warning: "The Bay is every day becoming more polluted ... and at any moment an epidemic may result."[9] Typhoid, cholera, and other waterborne diseases afflicted people in nearly every major Great Lakes city. But according to a 1982 report on "human occupation" of the region, "The faint voices of the few Public Health workers who expressed concern were effectively drowned in the roar of factories and rumble of the wheels of progress." After all, the report concluded, "The goals of the new cities were jobs and growth, not water quality."[10]

But the filth and stench in the waters of Great Lakes towns could be seen, tasted, and smelled, so the problem could not be ignored forever. In 1909 the United States and Britain, on behalf of Canada, signed the Boundary Waters Treaty declaring that the boundary waters between the two nations, including the Great Lakes, "shall not be polluted on either side to the injury of health or property on the other."[11] To administer the treaty, the governments established the International Joint Commission. The IJC, with commissioners appointed directly by the U.S. president and Canadian prime minister, would meet regularly to discuss ways of ensuring clean lakes and report back to the governments. But it was to have no special enforcement power of its own.

The first official declaration that the lakes were suffering from human activities came on August 1, 1912. In a joint letter to the International Joint Commission, the governments of Canada and the United States asked the commission to examine and report on "the facts and circumstances connected with the pollution of the boundary waters."[12]

At the conclusion of its study the commission described the entire stretch of boundary waters, including the St. Marys River, St. Clair River, Detroit River, Niagara River, and the St. Lawrence River from Lake Ontario to Cornwall, as "polluted to such an extent which renders the water in its unpurified state unfit for drinking purposes."[13] The source of the pollution, it reported, was the sewage and storm flows from the cities and towns and the sewage from ships. The commission argued that pollution control facilities were essential, and concluded: "It is feasible and practicable, without imposing an unreasonable burden upon the offending communities, to prevent or remedy pollution."[14] The commission reasoned that the two countries could maintain the health of the water by treating sewage before it reached the lakes.

The International Joint Commission's condemnation of the quality of Great Lakes waters spurred some municipalities to begin constructing primary sewage-treatment works to deal with the most obvious sewage-pollution sources. Many jurisdictions decided, however, that instead of building the necessary treatment works they would undertake the less costly measure of chlorinating drinking water and building water intakes farther out into the lakes. In his novel *In the Skin of a Lion,* Michael Ondaatje described the building of one such intake at the east end of Toronto.

> Work continues. The grunt into hard clay. The wet slap. Men burning rock and shattering it wherever they come across it. Filling hundreds of barrels with liquid mud and hauling them out of the tunnel. In the east end of the city a tunnel is being built out under the lake in order to lay intake pipes for the new waterworks.
>
> It is 1930.... the muckers move forward with their picks and shovels ... during this mad scheme by Commissioner Harris to collect lake water 3,300 yards out in the lake?[15]

Building new intake pipes did not get to the source of the problem – the dumping of sewage itself – and much of the flow of wastes to the lakes continued. Although the measures did largely eliminate typhoid and other waterborne diseases, the local government had avoided a long-term solution to the pollution of the lakes in favour of less costly temporary measures. They had missed an important opportunity to deal firmly and completely with the sewage-pollution problem, and the long-term ecological implications of their decision would become apparent only decades later.

THE NEW INDUSTRIAL ONSLAUGHT

While sewage was the major focus of the International Joint Commission report, the commission also warned of a larger and potentially more significant threat: pollution from the rapidly expanding industrial sector. The commission singled out wastes from the manufacturing and chemical industries for special concern but, despite its warning, still declared its faith in the dilution approach to pollution. The report concluded: "The immensity of the boundary waters, and their consequent capacity for dilution, will probably for some time to come prevent pollution from this source other than sawmill and pulp mill wastes becoming an international question."[16]

It would be almost thirty years before the commission began examining this question and meanwhile Great Lakes industry continued to expand. The commission itself remarked on the inevitability of regional growth in its 1918 report.

> The directness of the water route from the Atlantic Ocean to the head of Lake Superior, the adaptation to water carriage of the freight borne by the lake boats, and the cheapness with which this freight can be transported by them, the completion of the barge canal from Buffalo

> to New York, the proposed enlargement of the Welland and other canals along the boundary rivers permitting the passage of vessels of 28 or 29 feet draft from the ocean to the heart of the continent, the future settlement of the great wheat belt of Canada, the fringe only of which has been touched, and the possible utilization of the 3,375,000 dependable horsepower of the boundary rivers, render the conclusion inevitable that the commerce and shipping on these waters and the wealth, the industries, and the population along their banks must in the near future reach dimensions far exceeding their present attainment, and may ultimately far surpass any area of similar extent in the world.[17]

By 1940 the eight Great Lakes states accounted for more than two-thirds of total U.S. industrial production. The automobile had replaced the horse-drawn buggy and commercial air travel had taken its first flights. World War II added to the boom, with steelmaking still at the heart of expansion. Northwest of Lake Superior, mining companies found new ore reserves, increasing the flow of this precious war material down the lakes. It was necessary to expand the Sault locks to accommodate the larger ships moving Great Lakes resources, and improvements had to be made to the Welland Canal.

Mining, manufacturing, steelworking, and rubber production all concentrated their efforts on producing ammunition, planes, tanks, and ships. In 1944, 384 freighters were on the lakes. The ships, as author Harlan Hatcher described them, moved "back and forth with almost clocklike regularity between the source harbors on Lake Superior and the manufacturing and outlet ports on Lake Erie."[18] Each ship's load was equivalent to about 390 freight cars, with iron ore alone accounting for over ninety million tons of the cargo. The Great Lakes region was buzzing with activity.

Wetlands, once again, became a casualty. During the 1930s and 1940s, drainage ditches were cut through the Holland Marsh, an extensive wetland twenty miles north of Toronto, and one of the largest inland wetlands in southern Ontario. Within a few short years, the thick black soil of the marsh was producing millions of pounds of onions, lettuce, beets, and cauliflower. The production was so remarkable that the area was described as "central Canada's vegetable crisper."

World War II also brought about a dramatic increase in the production of synthetic chemicals and associated products. Nylon, the "miracle fibre," was first marketed by its discoverers, Du Pont, in 1939 and the advent of war expanded its uses significantly. Lee Niedringhaus Davis, a chronicler of chemical industry history, explained the impact of the war on nylon production:

> When Pearl Harbor was bombed in December 1941, Du Pont moved immediately to revamp the nylon plants, shifting from hosiery to the heavier yarns required by the military. By February of the next year, reams of nylon parachute shrouds, tow ropes, and tire cord (for air force planes and carriers) poured out of the Du Pont works.[19]

During the war deadly chemical poisons such as DDT also burst onto the market. Much fanfare accompanied their introduction; the manufacturers boasted that their new products had almost magical qualities. The volume of chemical production in plants along the Niagara and St. Clair rivers and at the Dow Chemical Company in Midland, Michigan, expanded greatly. By 1947 U.S. companies were producing over 124 million tons of synthetic pesticides. The synthetic chemical age was in full swing.

The frantic industrial activity in the service of war swept aside concerns about pollution and the environment. "No one, it seemed, had time to worry about the state of the Great Lakes," concluded Wayland Swain, a Great Lakes research scientist.[20] Predictably, the lakes, landscape, and wildlife suffered. The Buffalo Sewer Authority had reported in 1937 that on three occasions pollution had killed large numbers of fish in the Buffalo harbour area. There were fourteen other incidents of fish kills reported along the upper Niagara River in the years from 1943 to 1949. And the pollutants entering the lakes were becoming more and more diverse. Cyanides, phenols, copper and iron salts, and chlorine: all of these and still other materials were discolouring the waters and destroying life. Wayland Swain described the pollution of the lakes as "a mounting tide of acids, waste oils, fugitive greases, dangerous chemicals, paper sludge and general trash."[21]

By 1946 the two national governments were once again forced to respond. The severity of pollution problems stemming from both expanding industrial operations and the sewage of increasing populations prompted them to ask the International Joint Commission to investigate the lower Great Lakes and see if the Boundary Waters Treaty was being violated.

The commission completed its investigation in 1951 and its conclusion was straightforward: pollution of the boundary waters "is taking place to an extent which is injurious to health and property." The report noted specifically, "There is progressive overall deterioration of the waters of the Lower St. Marys River, the St. Clair River, the Detroit River and the upper Niagara River."[22]

This time the pollution problems were materially different than those found in 1912. The earlier investigation was concerned only with bacterial pollution caused by domestic sewage. The industrial processes then in use, said the commission, "did not discharge waste products in sufficient quantities to affect seriously the condition of the waters."[23] By 1951 the situation had dramatically changed. "Although the matter of industrial waste was of little or no concern in 1913, it is today a major problem," the Commission stated.[24] The 1951 report underscored the accuracy of the earlier prophecy that industrial wastes would in the future endanger the health of the Great Lakes.

The problem was enormous. Each day private companies were discharging over two billion U.S. gallons of wastes into the boundary waters, including a sizeable volume of hazardous chemicals. The pollutants included thirteen thousand pounds of phenols, eight thousand pounds of cyanides, twenty-five thousand pounds of ammonium compounds, and large quantities of oils and suspended solids of all types.

But, although the concerns over pollution had dramatically shifted from the former narrow focus on bacteriological problems, the earlier problems created by sewage disposal had not disappeared. Most enterprises, it seemed, still thought of the Great Lakes system as a convenient sewer that could suck up and take out of sight and thought whatever was not needed.

Some four million people continued to use the lakes as a source of drinking water and the fear grew that a particularly strong concentration of pollution would undermine proper treatment of this source. The IJC warned, "The precautionary measures that can be taken at water supply intakes are incapable of insuring that such 'slugs' of polluted water will not enter the water supplies and cause injury to the health of many water users."[25]

The continual discharge of untreated or, at best, partially treated municipal waste caused bacterial contamination, so much so that, alarmingly, the IJC study found that on average bacterial concentrations in Great Lakes waters were four to five times greater than they had been in 1912. Users of public beaches were the first to suffer, as municipalities started to close down beaches in industrial and large urban areas. Local authorities closed down the popular Bay Beach in Green Bay in 1943 as a result of bacterial pollution. Swimming at Burlington Beach on the western end of Lake Ontario had been stopped as early as 1924. Tourist operators in the vicinity of the beaches complained about loss of business, and beachfront property values began to drop.

But the beaches were just the front line. Inadequate municipal and industrial water-pollution control had another, more serious effect. Decomposing sewage and organic industrial effluents deplete a lake's valuable oxygen content, and the International Joint Commission study found evidence that the Great Lakes were beginning to suffer from this phenomenon. It appeared that the combined effects of municipal and industrial wastes were destroying the lakes.

To address these problems the commission recommended the construction of pollution control facilities and stressed the urgency of establishing primary sewage treatment, including waste sedimentation and disinfection. But the commission also argued that these measures alone were not sufficient: they should be followed by more efficient secondary treatment. The technology necessary to control the waste is available and affordable for both industry and municipalities, the commission said. The cost of both primary and secondary municipal waste treatment would be $76.5 million in the U.S. and $25 million in Canada. Industrial installation of pollution control facilities, the commission suggested, would cost an additional $22.6 million in the United States and $3.45 million in Canada. The commission also recommended that municipalities tackle the thorny problem of combined sanitary and storm sewers. During storms large amounts of waste from combined sewers bypassed the existing sewage-treatment facilities.

Concluding its recommendations, the commission stated: "The costs of the necessary remedial measures should be borne by the municipalities, industries, vessel owners and others responsible for the pollution."[26] Those who created pollution through their business practices and manufacturing were the most logical target for controlling wastes.

The International Joint Commission did not have the power to force the implementation of pollution control measures. The IJC could only make recommendations, and the two national governments would have to carry them out. In a final note of warning that proved significant for decades after, the commission remarked, "Past experience has shown that constant effort and attention by regulatory authorities are needed if existing pollution is to be controlled and new pollution is to be prevented."[27]

The national governments were reluctant to do anything that might slow the growth of industry and hamper suburban and urban expansion. The war years and after saw a technological and industrial boom. Business and industry focused their attention not on controlling wastes but on production of more, faster, and increasingly sophisticated products such as cars, washing machines, electric mixers, and televisions. Not surprisingly, in this spiral of production and consumption both the public and politicians largely forgot about considering the environmental costs.

THE DEMISE OF THE NATIVE FISH

The ailing health of Great Lakes waters did not deter fishers, but over a period of several decades they too had to cope with changing conditions.

The turn of the century was a period of intense fishing activity. More commercial fishers than ever before were setting out nets in the waters of the Great Lakes and hauling in a variety of species, depending on location. Catch statistics from 1899 reveal that fishers on the upper three lakes were mostly bringing in lake trout, whitefish, lake herring, and walleye. On Lake Michigan and Lake Huron they were also catching yellow perch. On Lake Erie, the most prolific of the lakes for fish production, the catch was lake herring, blue pike, carp, yellow perch, sauger, whitefish, and walleye. In the deep waters of Lake Ontario, never able to support commercial fishing as successfully as the other lakes, the prime catch by weight was lake herring.

But by the early decades of the twentieth century important species had suffered declines. Lake herring, the largest single catch on Lake Erie in 1899, had almost totally disappeared by the 1920s, falling from a catch of almost forty million pounds in 1899 to less than five hundred thousand pounds thirty years later. On Lake Michigan catch totals for sturgeon went from over five million pounds in 1885 to less than one hundred thousand pounds in 1916. Important whitefish areas in the north part of Lake Huron became less and less productive. By the late 1930s the whitefish catch totals there had declined to less than half the previous levels. Walleye catches on Lake Superior plummeted from 248,000 pounds in 1885 to 121,000 pounds by 1925.

To many fishers it became obvious that pollution was disturbing their livelihoods. During the First World War fishers in Saginaw Bay began to receive complaints about the taste of their catch and they hired Professor Herbert W. Emerson, a biochemist at the University of Michigan, to investigate the cause of the problem. He and another scientist, noting the bad odour and taste of Bay City drinking water, traced the cause of the tainted fish and water to the discharge of chemical waste from the Dow Chemical plant at Midland, Michigan, forty miles upstream on the Saginaw River. Dow was dumping its chemical wastes directly into the river, and dichlorobenzol, a heavy, clear, oily liquid, was both killing fish and causing taste and odour problems downstream. In Saginaw Bay the chemicals stunted the growth of the fish.

In April 1917 the Michigan attorney-general issued an injunction against Dow to stop its pollution and the company responded by building a settling basin for the waste. Although an overflow continued to pollute the river, the government decided not to impose more stringent controls. A fisheries scientist who studied the problem explained why: "Since Dow had military contracts, the several governments having jurisdiction were not inclined to be more restrictive to it during the war."[28]

On Green Bay the Wisconsin Conservation Commission and State Board of Health issued a special report in 1927 decrying the pollution by pulp and paper companies on the Fox River. They wrote:

> Even the more resistant fishes, which inhabit the lower reaches of the river, cannot withstand the combined effects of extensive pollution, low stream flow and high water temperatures which frequently exist during the latter part of the summer and early fall. The death of a large number of fish in the section of the river from Wrightstown to Green Bay had become an almost annual occurrence.[29]

At the same time other dramatic changes occurred in the Great Lakes fishery. One of the most noticeable was a shift in size of fish, from large, older fish to small, younger ones. Mature fish had originally dominated the population of the lakes. At one time fish weighing over ten pounds had comprised over half of the total weight of fish caught. With the rapid exploitation of the fishery in the twentieth century, the average size of caught fish went down and down, until the majority caught weighed less than a pound. Both abundance and size of fish were being altered.

The decline of the valued fish stocks had produced what author Tom Kuchenberg called "the management fad of the nineteenth century – the hatchery."[30] These operations, Kuchenberg stated, "were originally promoted by private entrepreneurs but quickly caught the fancy of governments. Fish were disappearing and artificial augmentation of stocks was a welcome alternative to restrictions on the fishery."[31] Governments on both sides of the international border promoted hatcheries and between 1867 and 1920 eighteen were constructed on or near Lake Erie alone.

Some fishery experts, deciding that hatcheries were not the answer to the fishery problems, worked to enact regulations to reduce fishing intensity. But co-ordination and co-operation among fishers and fishery officials in the eight states and Ontario proved extremely difficult. By 1905 some regulations were in effect, but stringent restrictions seemed too much to hope for. Instead, officials searched for quick and easy solutions. Their belief in technology and the human ability to manipulate the environment was apparently as entrenched as it was in other sectors of society.

Introduction of new species, fishery officials decided, was the logical answer to the demise of native fish. "The U.S. Fish Commission seemed bent on trying everything everywhere in hope that something would work," one scientist said.[32] The Canadian government followed the trend and, in addition to other introductions, began planting carp in Ontario between 1880 and 1893 as a means of "furnishing in the future a cheap article of food."[33]

Within nineteen years the Ontario Department of Fish and Game concluded that the successful introduction of carp had been a fisheries nightmare. The fish severely disrupted the shallow-water environment by uprooting aquatic plants, and stirred bottom sediments with their thrashing bodies. In addition, their competitiveness with native fish forced the government to concede that "the promiscuous introduction of carp on this continent has been attended with nothing but evil results."[34]

Michigan fishery officials also brought in smelt, with a first planting in the St. Marys River in 1906. They intended the smelt as a food source for hatchery-released salmon but neither smelt nor salmon survived. A remarkably successful second effort to establish smelt was made in Crystal Lake, Michigan, in 1912; spawning of the smelt was observed in 1918. Moving downstream from Crystal Lake, the smelt found their way to Lake Michigan and five years later local fishers were making commercial catches. The fish made their way to Green Bay in 1924 and were found throughout the lake by 1936. Commercial smelt production on Lake Michigan rose from 86,000 pounds in 1931 to over 4.2 million pounds in 1940.

By the mid-1930s smelt were present in all the Great Lakes, although the first significant commercial catches of smelt did not occur in lakes Erie, Ontario, and Superior until the early 1950s. But the presence of smelt did little to add to the stability of the fishery. Once introduced, their populations fluctuated dramatically. Disease struck Lake Huron and Lake Michigan smelt in the winter of 1942-43, and in 1944 fishers caught only five thousand pounds. In the following year the catch was a hundred thousand pounds, and by 1948 it had climbed to over a million. Ten years later the weight of the catch went over nine million pounds.

The instability of the smelt population was ecologically and economically unsettling. As Tom Kuchenberg remarked, "If the results of the carp planting frenzy was a strong indication that it was not wise to disturb the species balance, the smelt experience was final confirmation."[35]

ALEWIVES AND LAMPREYS: THE MARINE INVADERS

Overfishing, the destruction of spawning habitat, the reduction in the quality of the water, and the purposeful introduction of non-native species were all having an effect on the native fish populations. But the death knell for much of the native Great Lakes fishery was the unintentional introduction into the lakes of marine invaders, which spread rapidly throughout the lakes in the early years of the twentieth century.

The alewife, a small salt-water fish of the herring family, was the first of these invaders. Alewives most likely reached Lake Ontario via the Erie Canal, but were probably not able to thrive in the presence of the predatory Atlantic salmon. By 1873 salmon numbers in Lake Ontario had declined greatly, allowing the alewife to flourish and to eventually reach Lake Erie, probably using the Welland Canal to circumvent the Niagara Falls barrier.

Although alewives found favourable conditions in Lake Ontario, it was not so for Lake Erie, which as a shallow lake was too cold in winter to allow for prolific growth. Lake Huron was more to their liking. They first appeared there in 1933 and spread quickly, helped in this by the coincident decimation of their major predators, lake trout and whitefish, by another marine invader, the lamprey. Without predators to control it, the alewife population exploded. In 1949 alewives reached Lake Michigan and by the winter of 1956-57 fishers in the southern part of the lake were complaining that the fish was infesting their chub gill nets. In 1957 fishers caught 220,000 pounds of alewife in the lake. Three years later they hauled in 2.4 million pounds and in 1966 an unbelievable 29 million pounds. Alewives also reached Lake Superior by 1953, but did not flourish in that large, deep, cold lake.

Fishers regarded the alewife as a junk fish, useless. "They are not good to eat, and there is no sport in catching them," stated a 1968 government report about the fish.[36] Efforts to find a commercial market for them, as animal food, were only partially successful. By competing for food supply, they crowded out more desirable species. Biologist Stanford Smith argued that alewives had limited the populations of lake herring, emerald shiner, deepwater cisco, and deepwater sculpin. He noted, "Food competition appears to be the mechanism by which the alewife dominates other species. The alewife's ability to reduce or eliminate all large plankton from a lake may deprive other, larger species of the food necessary for their survival."[37] By reducing the numbers of the deepwater planktivores (cisco and sculpin), alewives may also have reduced the numbers of lake trout, which fed on these species.

Alewives may have brought about the demise of the native fish species or they may have simply filled a niche vacated by them. What is not open to question, however, is that by the early 1960s, alewives were a dominant feature of the fish fauna of the Great Lakes. The large quantities caught were used almost exclusively for fertilizer and pet food. By 1965, the Michigan Department of Natural Resources reported: *"The alewife presently makes up over 90% of the weight of all fish present in the Great Lakes."* Officials at that time declared, "It is so numerous that it now poses a threat to the survival of all species spawning within the Great Lakes."[38] The 1968 government report said that alewives had become "pests mindful of the great locust plagues recorded in history" and that: "Worst of all they move in enormous schools from the deeper recesses of the lakes, especially Lake Michigan, into inshore waters and die here by the millions – clogging water intakes and piling up in stinking masses on shore."[39]

Despite its destructive and disruptive influence on the native fish of the Great Lakes, the alewife does not receive top honors as the most destructive of the marine invaders. That award goes to the parasitic sea lamprey. Once established, it wreaked havoc on the native populations of lake trout and whitefish.

The lamprey, which looks like an eel, has a mouth in the shape of a round disc, with various bony, tooth-like appendages that attach themselves to other fish, rasping a hole in its prey's skin to suck off blood and body fluids. In a parasitic adult phase of its life, lasting from twelve to twenty-two months, the lamprey is extremely destructive, able to kill up to forty pounds of fish. After this phase the fish returns to streams to spawn and die, but the damage has been done. During their peak abundance in Lake Michigan, for instance, lampreys destroyed over five million pounds of fish annually. The fish that survive the attacks show circular scars where the lampreys had attached themselves. Don Misner, a Lake Erie commercial fisherman, recalled that fishers there were always catching whitefish that were scarred, which meant the fish were useless to the fishers. "You couldn't do anything with them," Misner said.[40]

The lampreys spread slowly after they were first observed in Lake Ontario in 1835 and did not make a noticeable impact on Lake Ontario fisheries until the late 1800s. By 1920 the lamprey population was on the increase, and by 1930 the fish were attacking lake trout, their favourite prey – evident in the dwindling commercial catch of that fish. In 1939 fishers harvested 285,000 pounds of trout in Lake Ontario; in 1944 they caught only 78,000; in 1949, 22,000; and in 1954, 8,000. In the early 1960s, New York State closed its fishery and the province of Ontario reduced quotas to a point large enough only to provide adequate samples for study.

By both swimming and attaching themselves to the hulls of ships, lampreys moved through the Welland Canal into Lake Erie, where they were first reported in 1921. Why lampreys took so long to make this journey remains a mystery, though modifications to the canal in 1919 may have removed a barrier to their migration. The impact of lampreys in Lake Erie was not as spectacular as in Lake Ontario. The tributary streams of Lake Erie contain more eroded sediment than those in Lake Ontario, so lampreys could not find the clean gravel beds they prefer for spawning. Nonetheless they moved up the Detroit River, through Lake St. Clair, and made it to the St. Clair River by 1930 and Lake Huron by 1933. Their arrival in the upper lakes had chilling consequences for native fish stocks. Within twelve years the commercial catch of lake trout had declined by over 60 per cent. By 1959 the landing of trout was almost non-existent. On Lake Michigan the lampreys' destruction of lake trout populations was even faster and more complete. In each of the first five years of the 1940s, fishers took over six million pounds of lake trout out of Lake Michigan. By 1953 the catch had completely disappeared.

Although the lampreys' favourite host was the lake trout, they attacked other fish as well. In summers the adult lamprey had moved to deepwater areas where they latched onto lake trout; when lake trout were eliminated, they began working on burbot and chub. In the fall lampreys moved shoreward and attacked whitefish and lake herring. In the fall of 1953, when lampreys were near peak abundance in Lake Michigan, a commercial fishing journal, *The Fisherman,* recorded a graphic illustration of their devastation:

> Lampreys in such vast numbers that the water literally teemed with them invaded Green Bay early this fall bringing fishing operations practically to a standstill. All important commercial species were attacked, and so great was the destruction that dead and dying fish littered the surface of the water. Fishermen of northern Green Bay, discouraged by dwindling catches, by the presence of large numbers of lampreys in their nets, and the necessity of discarding the high percentage of mutilated fish in the catches they did make, removed most of their gear from the water by the end of September.... Trap-net fishermen reported as high as 60 to 70% of their catch scarred and many dead fish when the nets were lifted.[41]

During the 1940s lampreys had also begun moving into Lake Superior, and by the fall of 1952 a spawn-taking crew of the Wisconsin Department of Natural Resources reported about one out of every ten fish had been scarred. Author Tom Kuchenberg summarized the lampreys' damaging influence in his 1979 book:

> In the space of 20 years this parasite devastated the fish stocks of the upper lakes and completely changed the species balance in their waters. It destroyed a large segment of the commercial fishing industry, and the resulting economic dislocations affected scores of small communities. It destroyed or severely reduced predator species, leading to an explosion of small native and colonizing species that disrupted the food chain. In short it produced a chaotic ecosystem and left aftershocks which persist to this day.[42]

By the late 1950s the fishery of the Great Lakes was vastly different from the one that the explorers and missionaries had boasted about in their correspondence to Europe. Modern enterprises – the clearing of the land, overfishing, construction of dams and canals, introduction of non-native species, and pollution in general – had disrupted the once abundant and diversified fish fauna. Sadly, the disruption of the fishery was only one symptom of the generally declining health of the lakes and the surrounding landscape.

FIVE

The Aches and Pains of Relentless Expansion

BY THE MID-POINT of the twentieth century, people everywhere were lauding the industrial development of the Great Lakes region. Senator Everett McKinley Dirksen of Illinois described the area as being "blessed with a soil unusual in its richness and abundance, with a splendid supply of fresh water and with superb mineral resources." The minerals were especially important to the region, he said, "for it is here that they are fashioned and fabricated into almost everything used anywhere in the world."[1]

Robert McLaughlin, an editor of Time-Life Books, further expanded on the industrial activity that was taking place around the Great Lakes at the beginning of the 1950s:

> Along the Great Lakes from Superior to Cleveland, stretches a thousand-mile belt of factories, mills, refineries, blast furnaces and machine tool plants. You name it – tractors or travelling cranes, copper or ketchup, limestone or lightbulbs, port or plate glass – the region produces it or grows it or claws it from the earth.[2]

The Chicago-Gary area produced more steel than all of France, Detroit built a third of U.S. automobiles, and the region as a whole produced three-quarters of U.S. construction machinery. The region generated an amazing one-fifth of the U.S. gross national product. By 1964 the value of industrial output on the U.S. side of Lake Erie alone was $17 billion.

On the Canadian side industrialization was not as extensive but it did hold a relatively more important position for the country. The region produced over one-half of Canada's gross national product. In 1964 the industrial output from the Canadian side of Lake Ontario was $2.8 billion. Toronto was a large, expanding city; Hamilton produced the major portion of the country's steel; and at Thunder Bay on Lake Superior grain elevators were filled to capacity with the agricultural gold of the prairies: wheat.

Completion of the St. Lawrence Seaway in 1959 gave a boost to industrial activity in the Great Lakes region. Previously, the small, shallow locks around the treacherous rapids of the St. Lawrence River had restricted the passage of large ships. Now the building of the seven larger seaway locks opened a new era in navigation. Ocean-going vessels flying the flags of distant nations could power their way up the St. Lawrence and into the Great Lakes. Other massive freighters previously confined to the lakes were able to make their way downstream. The International Joint Commission boasted in 1969: "The navigation facilities, channels, locks and harbours make the Great Lakes one of the most sophisticated water transportation systems in the world."[3] Ship traffic tonnages steadily increased and the cargo carried through the Welland Canal rose from 27.5 million tons in 1959 to 59.1 million tons in 1966.

The Seaway opened up new patterns of commodity flows on the Great Lakes. In the 1950s geologists found major new iron-ore deposits in Labrador, on the east coast of Canada, and companies could now ship the ore west to Great Lakes mills through the canals and locks of the Seaway – reversing the traditional direction of Great Lakes ore flow. The postwar economic boom was in full swing, and the Great Lakes region was flourishing. By 1966 one out of every three Canadians and one out of every eight U.S. citizens lived in the Great Lakes basin. In total, over thirty million people lived on or near the lakes.

But by the early 1960s it was also more obvious than ever before that a rider to these remarkable economic achievements was serious environmental damage. The evidence was everywhere and the mass-circulation press was catching onto it, telling stories of ruination accompanied by pictures of rotting algae and dead and dying fish.

The vast Great Lakes, once seemingly endless in their purity, were in deep trouble.

THE INESCAPABLE UNMISTAKABLE SYMPTOMS

"Lake Erie is dying" stated Canada's national news magazine, *Maclean's,* in November 1965.[4] In the same year *Time* declared Lake Erie "critically ill."[5] *Newsweek*'s head for a story about the Great Lakes was "The Dead Sea."[6] Other pronouncements about the death of Lake Erie and the visible destruction of the Great Lakes environment were widespread in the mass media of both Canada and the United States.

"The symptoms are there for all to see," the *Time* article said. The magazine pointed out, "Beaches that were once gleaming with white sand are covered with smelly greenish slime" and "The lake's prize fish – walleye, blue pike, yellow perch and whitefish – have all but disappeared." *Maclean's* called its story about Lake Erie "Death of a Great Lake":

> Lake Erie might have outlived the human race. Instead it's becoming a 10,000 square mile dead sea. By smothering it with pollution, man is making it an odorous, slime-covered graveyard. And we may have already passed the point of no return.[7]

Even the business press, with its admitted priorities of steady economic growth and healthy profit margins, could not avoid reporting on the visible damage. In 1966 *Canadian Business* commented, "Canada and the United States have used the Great Lakes indiscriminately for years for recreation, transportation and, tragically, municipal and industrial sewage disposal." The article asked, "Are the lakes to become Great Sewers, or can something still be done?"[8]

While the press was declaring the "death of Lake Erie" and the decline of the other lakes, the irony of the situation was that the Great Lakes were, in large part, too "rich in life." As many observers pointed out, one of the major reasons for the problem was an overabundance of algal life.

Nutrient-rich wastes from people and industry, when added to the lakes, had speeded the growth processes of certain algae, leading to a veritable explosion of algae life. The problem is, when algae die, they settle to the bottom of the lake, where they are decomposed by micro-organisms, a process that requires oxygen. When more and more of the dead material drifted to the bottom, the decomposing micro-organisms used up more and more of the lakes' precious oxygen. Trout, whitefish, and aquatic insects, which like the rest of us need oxygen to survive, could not live in the oxygen-depleted waters. So the lakes were, strangely enough, dying from an overabundance of life.

The first concrete sign of this problem was the disappearance of the mayfly in 1953. Every spring, for thousands of years, adult mayflies had emerged from their larval existence in the waters of the lake. They had danced around the light standards of lakeshore towns like moths and had littered the pavement below with their carcasses. Their 1953 disappearance proved to be temporary – they were seen again in subsequent years, although their numbers had declined – but by 1956 the adult mayfly had gone forever, eliminated by a depletion of oxygen. Without oxygen, their larvae could not survive in the bottom sediments of Lake Erie, where they had once been so common.

But the mayfly was not all that disappeared, given the supremely logical cause-and-effect cycle of nature. Mayflies provided an important food source for fish, especially the yellow pickerel, or walleye, commonly found in the western half of the lake, and blue pike, found in the eastern half. These fish had long been the mainstay of the world's biggest freshwater fishery. In 1936, U.S. fishers landed close to twenty million pounds of the prized blue pike and twenty years later the fish were still being caught in substantial quantities. Canadian fishers netted over forty-four million pounds of fish in 1956, about half of it blue pike and yellow pickerel, worth about $5,500,000. That was the last good year for blue pike. By the early 1960s the fish was extinct, never again to be seen in Lake Erie or in any other waters. At the same time the pickerel was also endangered: during the first four years of the 1960s, pickerel catches dropped by over 90 per cent.

The demise of the mayfly had in part led to the disappearance of the blue pike and the decline of the pickerel. But other factors also played a part. The lack of oxygen undoubtedly caused the suffocation of some blue pike and pickerel eggs as well as reducing insect food. Other contributing factors were overfishing, siltation of spawning beds, and toxic contamination.

The reduction in numbers of valuable fish was a blow to the Lake Erie commercial fishery. The earnings of fishers had been the mainstay of the economy for many shoreside towns. Now many fishers had to sell their boats and look for other work. Before the extinction of the blue pike there had been about three thousand Canadian fishers working Lake Erie. Ten years later only half of them were left. The U.S. side was even harder hit. Don Misner, who fished commercially out of Port Dover and proudly admitted to having been a fisherman all his life, saw a big fishery on the U.S. south shore dwindle down to being "almost non-existent." He said, "People who couldn't adapt to the changes in the fish species simply got out of the business."[9]

Along with this the value of Great Lakes property plummeted. Sitting mournfully and surveying his stack of boats and the empty sands of a Lake Erie beach, a fifty-eight-year-old boat rental operator, Forrest Dadlow, told his story to a *Maclean's* reporter in the early 1960s:

> I bought this place and 24 boats in 1958, figuring I could get a good living out of it, like the old guy who had it for years. But look now – high summer, bang in the middle of the school holidays and I'm lucky if I rent two boats a day at two bucks a time. Eight, nine years back the sport fishermen would have rented all these boats by Friday night. Then they'd be out all weekend and come back with a whole mess of big beautiful blue pickerel. Now there's only sheepshead and carp. So what have you got? No fish, no fishermen – no beach, no people.[10]

Lake Erie was "dying" and it was not the only Great Lake that was showing signs of serious environmental illness. Cladophora, a slimy, filamentous, green algae that grows on rocks, dock pillars, and breakwalls, became excessively abundant in the nutrient-enriched waters of Lake Michigan. Small clumps of cladophora had thrived on the southern edge of the lake for years and by the early 1960s had grown to several feet long. In periods of heavy wave action, the cladophora broke loose from the substrata they grew on and washed ashore to litter beaches in slimy windrows. The floating heaps of green plants clogged water-intake screens and interfered with swimming. The decaying cladophora gave off a putrid odour and provided an ideal breeding place for flies and other insects.

While the buildup of the cladophora forced the closure of beaches on both Lake Michigan and Lake Ontario, other beaches around the Great Lakes were closed because of the continued threat to swimmers posed by bacterially spread disease. Bacteria in human wastes are easily destroyed by disinfection during standard sewage-treatment processes, but a great deal of raw sewage continued to reach the lakes untreated, and authorities had no choice but to close beaches to swimming.

Industrial wastes also continued to take their toll on all the lakes. In the mid-years of the 1960s municipal and industrial waste made up an astonishing 95 per cent of the flow in the Indiana ship canal at the foot of Lake Michigan. Author Tom Kuchenberg warned that the pollution mixture was "so nightmarish" that it alone could lead to an irreversible deterioration of the lake.[11]

THE 1960S: EARLY WARNING SIGNS

By 1960 three centuries of European activities had greatly changed the Great Lakes and the surrounding environment. Many of the alterations that had begun in the previous century continued apace, almost unnoticed in the day to day progression of history. More and more land throughout the basin was paved over – a trend accelerated by the shift towards suburban development. In Wisconsin, for example, there was a 27 per cent increase in urban development between 1958 and 1967. Additional woodlands were cut down and wetlands were filled up to make way for the needs of expanding industry and the growth of large urban centres. The marshes of Michigan's Raisin River were lost in the process of providing land for factories and a waste disposal site.

Marsh after marsh disappeared. The rich organic soil of wetland areas prompted many farmers to drain them and convert them to crop production. In areas of clay soils, agriculturalists installed drainage projects to move the spring accumulation of water quickly from fields. The technique increased crop yield, but damaged the natural cycle of water flow in the rivers and lakes.

Farming became increasingly specialized, with the shift away from the family farm to larger farms and more mechanized farming methods. The old integrated or mixed farm, composed of a variety of animals and crops and land use, was gradually replaced by industrial, monocultural farming. Huge feedlots and single-crop farms became the norm. Farming became increasingly dependent upon the use of expensive fertilizers and chemical biocides. The alarming thing in this trend was the fact that many insects had already developed resistance to the sprays. By 1960 over 120 insects had been labelled as resistant to various chemical controls.

In the forests, improved fire control and an end to the logging era had allowed many of the cutover and burned lands to be regenerated during the early years of the twentieth century. But the composition of species in the regenerating forests was vastly different from the original. Aspen, a quick-growing species previously thought of as a weed tree, proliferated in the cutover and burned lands, forcing the development of new ideas for its use. Early in the century the forest industry began using it as a filler for packing material. After World War II, the industry developed a new pulping technique, which meant aspen could be used for paper production.

Pulp and paper production increasingly became dominant in the Great Lakes forest industry. By 1923 Green Bay had the distinction of being the world's largest producer of toilet tissue. As one local author noted, "Its daily output of 200 tons was enough to wrap twice around the equator."[12] The Fox River, which drains into Green Bay, became the site of the largest concentration of pulp mills in the world. Other mills sprang up along the north shore of Lake Superior in the towns of Marathon and Terrace Bay. The paper companies had their own priorities and their own peculiar environmental concerns: for one thing, the impact of insects that periodically ravaged portions of their valuable forests before they could get to them. Eventually this worry led to yet another cycle of chemical controls.

On the forest floor the wildlife populations continued to change significantly. The centre of the Great Lakes moose population moved steadily northward. In 1918 the wolverine vanished from its last Great Lakes stronghold in Minnesota. Other fur-bearing animals, such as marten, fisher, and lynx, became extremely scarce. But environmental changes did not adversely affect all wildlife. Deer still flourished in the areas of second-growth forest, and beavers, which had been driven almost to extinction by the end of the last century, rebounded spectacularly in the twentieth. They found abundant quantities of their favourite foods, aspen and birch.

In the twentieth century modern enterprises also caused significant alterations in the flows and quantities of water in the lakes. Expanding industry required energy sources to run its machinery. Companies dammed up many rivers, especially in Ontario, to increase hydro-electric production. Two such projects in northwestern Ontario reversed the flow of water in river systems to move south into the Great Lakes rather than north to Hudson Bay – diversions that added an average of 5,400 cubic feet per second of water to Lake Superior. There were also more everyday developments that had similarly dramatic effects: more and more people – and more and more private firms – were withdrawing water from the lakes for drinking and domestic, agricultural, or industrial use. This use was placing a new and unprecedented strain on water flow and supply.

By the middle years of the twentieth century the stresses on the Great Lakes environment were acute. Scientists, the media, and concerned citizens in both countries could easily identify a short list of environmental interventions – massive dumping of municipal and industrial wastes, or the invasion and introduction of exotic fish species – that were clearly detrimental to the ecosystem. Other activities – the clearing of forest land, the draining of wetlands, and the use of synthetic organic chemicals – had caused environmental alterations that observers found more difficult to detect. Nonetheless, the interaction of a variety of stresses and their cumulative impact had visibly damaged the Great Lakes environment by the early 1960s.

The severity of the problems produced a catalogue of bizarre phenomena. The weeds in Rondeau Bay on the north shore of Lake Erie became so dense that they looked "like a field of wheat" and an aquatic weed cutter was purchased to fight back the growth. The Cuyahoga River running through Cleveland was so clogged with oils and greases that it caught fire in 1969. The city had to build a fire wall and declare the river a fire hazard, leading *Esquire* magazine to bestow its "Dubious Achievement Award" on the unfortunate waterway. In March 1967 a deadly combination of cold weather and industrial pollution killed five thousand ducks along the Detroit River. Wood fibres, chips, pulp-paper mats, and oil slicks clogged the St. Marys River. Oil slicks and discoloured water were common on the Niagara River, while thousands still flocked to marvel at the falls. In January 1967 a worker's acetylene torch accidentally ignited the oils on the Buffalo River, a tributary of the Niagara. Flames leaped high into the air, burning pilings for a bridge and melting glass fixtures thirty feet above the surface of the water.

Rivers throughout the region were gaining a new but not necessarily welcome look. Hardest hit perhaps was the Detroit River, with its rainbow assortment of pollution. The discharge from the Detroit sewage-treatment plant caused a brown discoloration of the river, complete with accompanying odours. The Rouge River, entering the Detroit River at Zug Island, added a new batch of colours to the river water, almost living up to the promise of its name. Further downstream, industrial wastes from the Great Lakes Steel Corporation contributed a reddish discoloration. Along the U.S. shore Wyandotte

Chemicals Corporation added a white discharge. On the Canadian side, the absence of sewage-treatment facilities in Windsor tossed further unsavory solids into the concoction. The nearby Allied Chemical Canada Ltd. plant added its own white effluent.

Floating suds and dead and dying fish were everywhere around the lakes. Casey Burko, a veteran reporter for the *Chicago Tribune,* expressed the feeling of many who cared for the once Great Lakes:

> It was not hard to believe something terrible was happening if you stood on the banks of Cleveland's Cuyahoga River, a river that burst into flames occasionally, and saw thick mats of oil and grease ooze past like a gooey glacier. Or the Indiana Harbor Canal that flowed like a melted chocolate bar past the oil refineries and steel mills near Gary, Indiana. Or the Rouge River in Detroit that was as red as some of the fire-engine colored cars that rolled off the assembly lines at Ford Motor Co., which dumped 100,000 gallons of sulphuric acid pickle liquor into the river each day. It was impossible to look upon such environmental ruin without wondering: 'How did this happen?'[13]

Burko answered his own rhetorical question. "It was the American way in 1967," he said. But what he called the American way in 1967 had also been, for over three hundred years, the European way in the Great Lakes basin: with little or no regard for the natural surroundings, human activities had been guided by an abiding belief in the human ability to master and control the environment, and an unwavering quest for material prosperity.

The modern, capitalist exploitation of the Great Lakes region had produced incredible wealth and material well-being, but it had done so by ignoring the crucial foundations of all healthy development: the environment. This failure was obvious enough in 1912, when the International Joint Commission first studied pollution problems on the lakes and warned that the waters of the basin could not serve as the perpetual dumping grounds for human and industrial wastes.

The failure became all the more notable with the rapid depletion of fish stocks, the introduction of non-native fish, and the disruption of spawning areas. Still, the practices continued – it was difficult if not impossible to stem the "tide of progress" – and the forces driving the growth of industry and urban development were powerful. The early warning signs went largely unheeded and by 1960, for those few who cared, the lakes and the surrounding landscape seemed well on the way to irreversible ruin.

PART TWO

SIX

The Public Push for Environmental Responsibility

THERE'S A TRIP I like to make in the spring, to Long Point Provincial Park and Point Pelee National Park on the north shore of Lake Erie to watch the spectacle of the spring bird migration. On one visit, after walking along the sandy beach of Long Point I stopped to sit for a rest on a wave-washed tree anchored on shore. With my camera beside me and binoculars around my neck I gazed out over the lake, watching for unusual waterfowl or gulls. But as I scanned the water my thoughts turned from the birds to the writing of this book. Sitting there on shore it was hard to fathom the enormity of the damage done to a body of water as immense as Lake Erie, the damage done to the whole lake chain, the whole region. I recalled what I knew, the facts and figures, the horror stories, the history. I felt sad but I felt something else as well: the sense of pleasure and peacefulness you get from experiencing a lake's natural beauty – the ceaseless activity, the amazing variety of life – the sense of harmony, of hope. I felt connected to that activity and life. And I understood more sharply than ever that as the human element in this ecosystem we have a very real choice to make: the choice not to destroy the Great Lakes environment.

STUDIES, REPORTS, AND THE FACTS ON PHOSPHORUS

By 1960 the ill health of Lake Erie had prompted the U.S. government to authorize its Department of Health, Education and Welfare to conduct a thirteen-million-dollar study of the problems. More than one hundred experts spent two years studying the lake and C.W. Northington, who headed the group, summarized their grim conclusion: "It is still of use for industrial needs, but to fishermen, commercial and sporting, and to swimmers, boaters, lakefront-property owners and nature lovers the answer is yes, Lake Erie is dying."[1]

To lakeside residents and regular visitors, the results of the study only confirmed what they could see for themselves. In the 1960s, protests over the filth in the waters of the area had intensified. One irate Cleveland resident wrote, "Our lake is a wastebasket for factories. It is unfit for fish to live in and for people to enjoy."[2] This sentiment was not unique. In Ohio over a million residents petitioned the state governor to initiate action to reverse the pollution of Lake Erie. An environmental consciousness was being born among the general public. Environmental writer Barry Commoner perhaps best captured the popular sentiment of the time: "We have grossly, irreversibly changed the biological character of the lake and have greatly reduced, now and for the foreseeable future, its value to man. Clearly we cannot continue on this course much longer."[3]

Despite confirming the obvious, Northington's Lake Erie study brought much needed political attention to the problem. Senator Gaylord Nelson, a Democrat from Wisconsin, declared, "If the tragedy of Lake Erie is repeated in the other Great Lakes – as it may well be – the great industrial cities of America would be the victims of the greatest natural resource disaster of modern times."[4]

Something had to be done, and in a now familiar response to Great Lakes problems, the governments of Canada and the United States once again called upon the International Joint Commission to investigate. In October 1964 they asked the commission to recommend remedial measures to halt the deterioration of the lakes. The solutions, said the governments, must be practical from the points of view of both economics and sanitation.

To conduct the study, the IJC established two boards composed of scientific experts: the International Lake Erie Water Pollution Board and the International Lake Ontario-St. Lawrence River Water Pollution Board. The IJC instructed these boards to determine the extent, character, and causes of the pollution problem. It would be five years before the two boards completed their final reports, but the importance of their early findings prompted the IJC to send an interim report to the governments only a year after the study began.

Submitted in December 1965, the IJC interim report concluded that although information was far from complete, what the boards had learned so far indicated that the situation, especially in Lake Erie, was "serious and deteriorating." The major problem was cited as "eutrophication": the process whereby a lake or river suffers from too much plant growth caused by a high concentration of nutrients, such as phosphates, resulting in a decrease in oxygen.[5]

Dr. Jack Vallentyne, a senior scientist at the Canada Department of Fisheries and Oceans, was a member of the IJC study team. He later recalled, "At the time, there was not conclusive proof of the factors that caused eutrophication."[6] European scientists were leading the way in the study of eutrophication and had identified phosphorus and nitrogen as the nutrients responsible for the problem. A chance meeting with one of the European scientists prompted Vallentyne to undertake dramatic demonstrations that illustrated the importance of reducing the inputs of phosphorus, the most easily controlled of the two elements.

Vallentyne collected water samples from Lake Erie and Lake Ontario. Back in his lab, he set up one sample of pure lake water. He enriched four other samples with equal amounts of sewage wastes that had gone through different types of treatment. He added raw sewage effluent to one mixture. He added sewage effluent from a good secondary sewage-treatment plant to a second. He mixed a third sample with effluent from a sewage-treatment plant that had removed phosphorus from the water. He added phosphorus to a fourth sample, in an amount equal to that taken out by the phosphorus removal plant. He incubated all the mixtures and, at the end of the test, filtered the contents of the samples to determine the amount of algal growth.

The experiments were simple enough, but they yielded graphic results. The sample of lake water with nothing added left only a little crud on the filter paper. The mixture of lake water and untreated sewage effluent left an ugly mass of material on the filter. The filter used for the mixture of water and wastes from the secondary sewage-treatment plant looked just as bad. The filter used for the sample of water mixed with effluent from a sewage plant that had removed the phosphate looked almost as clean as the one used for the lake water alone. The final sample, with the phosphate re-added, left just as much crud on the filter as those used for the samples with untreated sewage waste.

The scientific report to the IJC on the pollution of the lakes included photographs of Vallentyne's filters. His experiment clearly illustrated that the discharge of phosphorus-rich effluent from municipal sewage systems and some industrial facilities was not only one of the most damaging pollutants being discharged into the lakes – but also that it was one of the most controllable.

Between 50 and 70 per cent of the phosphorus in municipal sewage came from detergents. Most of the rest was from human excreta. The detergent manufacturing industry, dominated by Proctor and Gamble, Colgate-Palmolive, and Lever Brothers, had started using significant quantities of phosphorus in detergents in the 1940s. Phosphorus served as a "builder" in the detergents: it was to soften the water so that the active cleaning ingredient could do its job more effectively. By the late 1960s the inorganic phosphate content of detergents varied between 30 and 50 per cent of the weight of the product. Discharged down the drains of millions of households around the basin, phosphorus-laden suds added significantly to the eutrophication problem. In December 1965 the IJC interim report concluded that excessive inputs of these algae-stimulating nutrients had to be stopped.

> Although there is as yet no conclusive evidence that the removal of phosphates, one of the essential nutrients involved, from the wastes discharged would in itself reverse the effects of eutrophication, the Commission, on the advice of its technical advisers, is satisfied that such action would materially retard further deterioration in the quality of these waters and should be taken.[7]

A 1968 U.S. government report on the pollution of Lake Michigan was equally blunt in calling for measures to reverse the damage of pollution and eutrophication in particular: "Eutrophication is a threat now, to the usefulness of Lake Michigan and other lakes within the Basin; feasible methods exist for bringing this problem under control." These controls "need to be applied," said the report.[8]

By 1969 the Technical Advisory boards to the International Joint Commission on Great Lakes pollution had completed the studies begun in 1964. Their report, *Pollution of Lake Erie, Lake Ontario and the International Section of the St. Lawrence River,* was released in 1970. The results of the study were not encouraging. The report concluded, "The introduction and accumulation of untreated or partially treated wastes from tributaries, municipalities and industries has [sic] limited the legitimate uses of the Lakes, caused unfavourable biological changes, and destroyed much of the general satisfaction and enjoyment that we refer to as the Lake's contribution to the quality of life."[9]

The report cited pollution from heavily industrialized areas, along the St. Clair River, the Detroit River, the Maumee River, the Cuyahoga River and the regions of Buffalo / Niagara Falls, Rochester, and Hamilton / Toronto, as particularly damaging. Pollution had "induced adverse biological changes," the report stated. It also identified other problems, none too surprising to those who had been following ecological trends in the region: dramatic changes in fish populations; bacterial contamination along the shorelines; increased accumulation of dissolved solids and wastes in the lakes; increased water-treatment problems; impaired recreational and aesthetic value of the lakes; and destruction of wildlife.

In a straightforward manner, the IJC described the need for corrective action. "Adverse effects on the environment and on the 18 million people who live on or near the shorelines, will become more critical unless there is vigorous and concerted implementation of imperative remedial measures."[10]

CITIZEN ACTION: THE RUMBLE OF COMPLAINT

The 1970 report of the IJC provided further ammunition for a public increasingly conscious of the need for environmental protection. Public spokespersons began to challenge the historic chain of decisions and activities leading to the destruction of the environment.

All around the basin citizens initiated anti-pollution campaigns. Housewives to End Pollution, a Buffalo women's group, launched a campaign to force local supermarkets to list the phosphate content of soaps on the grocery store shelves. "We let it be known that we were willing and able to supply pressure tactics if our requests were not answered," Carol Kaltwasser told a conference on pollution of Lake Erie in June 1970.[11]

Pollution Probe, a Toronto-based environmental group, conducted one of its many creative exposés of local pollution problems, holding a highly authentic mock funeral for the Don River. A eulogy to the "dead river" noted that in 1830 the river had been so pure that four breweries used it to make beer. By 1969 the pollution load in the river consisted of twenty thousand pounds of suspended solids, and bacterial counts were found to be as high as sixty-one million for every hundred millilitres; bacterial counts for safe swimming were set at 2,400. A university chaplain expressed his hope for the future of the river, predicting that a new generation, conscious of the environment and pollution, would lead to a restoration of rivers and streams like the Don.

At Silver Bay, Minnesota, local people became engaged in what turned out to be one of the longest battles anywhere to stop lake pollution. Silver Bay, at the western end of Lake Superior, is the location of the iron-ore processing plant of the Reserve Mining Company. The company dug out taconite rock – which contains about 30 per cent iron ore – and then ground it up into fine particles smaller than baking flour. Powerful magnets separated the ore from undesirable silica and other rock materials, and machines compressed the extracted iron into pellets ready for shipping.

To extract the iron from taconite the mining company used large quantities of water to transport and wash the taconite powder. Fortunately for Reserve Mining, nearby Lake Superior was a bountiful source for the seven-hundred million gallons of fresh water needed every day. But this water, when Reserve Mining was finished with it, was not the clear, clean material originally pumped from the lake. Now diluted in the water was a staggering number of fine waste-rock particles. Each day, into the blue waters of Lake Superior, the company dumped a soup containing sixty-seven thousand tons of fine waste minerals.

In 1947 Reserve Mining applied for a permit from the State of Minnesota and the Army Corp. of Engineers to discharge its "tailing waters" into the lake. It got the permit despite the warnings of one official who stated, "It is my opinion that the deposition of the tailings of the waters of Lake Superior will be harmful to aquatic values, and may eventually cause a reduction in the supply of lake trout, whitefish, and lake herring by destroying their food supply and the covering of their spawning grounds by depositing silt from the tailings."[12]

By the late 1960s, long-time residents in the Silver Bay area had begun to notice changes in the lake, especially peculiar, large patches of water with a milky green hue. A team of scientists who investigated the problem concluded that mine tailings were discolouring the water over an extensive area at the western end of Lake Superior. The conditions had also greatly reduced bottom-feeding insects and other organisms and, alarmingly, water in Silver Bay was shown to contain aluminum, lead, copper, zinc, cadmium, and nickel in amounts considered harmful to aquatic life.

People in the area, understandably aroused by news of the tailing pollution, began to fight back. At a public meeting called to discuss the pollution, Walter Sve, a commercial fisherman from Split Rock, Minnesota, told government officials, "It was my patriotic duty to this great country of ours to come and make a statement.... We have this beautiful lake and it is being wrongfully misused."[13]

The rumble of complaint grew louder and federal and state government officials responded to the public outcry by organizing a three-day conference on the tailing problem. More than six hundred people attended and over one hundred citizens presented statements to the conference, held in May 1969. Among the groups to testify were the Sierra Club, the Wisconsin Wildlife Federation, the Izaak Walton League, the United Steelworkers of America Local 5296, the League of Women Voters, the Minnesota Federation of Labor, and the United Northern Sportsmen of Minnesota.

Other citizens organized themselves into chapters of the Save Lake Superior Association. Hundreds of people joined the group and its president, Arlene Lehto, expressed the prevailing sense of anger in a 1972 statement to an IJC hearing. Lake Superior is "renowned for its purity and clarity," she noted, but industry, "through a combination of political string-pulling, economic fear tactics, legal maneuvering and sheer gall, has circumvented justice and continues to pour a mountain of filth into this international body of water."[14]

At the same hearing Robert Peterson of the United Auto Workers declared, "We have long regarded Lake Superior as being our greatest natural resource. It disturbs us to see this sky blue water turned to murky brown."[15]

But the fight against the pollution of Lake Superior was not to be won at a single three-day conference or government hearing. Until well into the 1970s the Save Lake Superior Association studied the tailings problem, steadily building up evidence of how taconite tailings were seriously harming the lake. It found, for instance, that each day's batch of tailings dumped into the lake included 51,500 pounds of phosphorus. Public awareness had changed, but industrial practices had not.

Companies like Reserve Mining argued that they were in the

business of providing not only important products but also jobs, and that the loss of those jobs through environmental protection would only hurt workers. This led unionist Robert Peterson to conclude, "What we need is a new social ethic, which would affirm and implement the basic proposition that we have a right and a means to ensure both a wholesome environment and economic stability."

But the two government levels did little to change matters until finally, in 1977, the U.S. Environmental Protection Agency (EPA) announced an order to Reserve Mining to stop the dumping. The company responded by challenging the order, forcing the EPA in turn to launch a lawsuit. The Save Lake Superior Association and the Northern Environmental Council joined the government in the lawsuit against Reserve Mining. The trial lasted an amazing 146 days and included more than one hundred witnesses and nineteen thousand pages of transcript.

At the trial's conclusion, the presiding judge ruled that Reserve Mining had to halt dumping of tailings into the lake "because the discharge into the water substantially endangers the health of people who procure their drinking water from the western arm of Lake Superior, including the communities of Beaver Bay, Two Harbors, Cloquet and Duluth in Minnesota; and Superior in Wisconsin." He concluded his statement by noting that the company has "the economic and engineering capability to carry out an on-land disposal system."[16]

While it might have had the capability, the company did not have the inclination. Like many other enterprises caught in similar situations, Reserve Mining showed more interest in safeguarding its balance sheets and profit margins than in covering the unwelcome costs and disruptions involved in protecting the environment. Although the court decision imposed an immediate shutdown of the plant, the company appealed to a higher court, which ruled a few days later that the plant could remain open while further appeals were heard. Ten years would pass before the company finally completed the on-land disposal system.

In the meantime citizens had proved themselves willing to take on the challenge of protecting the Great Lakes environment. It was also apparent that owners and managers of corporations would vigorously resist what they saw as public encroachments on their right to carry on in the profitable manner to which they had become accustomed. But more and more, all across the continent, citizens were not so ready to let this business-as-usual approach prevail at the expense of the environment.

In recognition of the popular sentiment, the 1970s began with Earth Day, a celebration intended to instil an awareness of the fragility of the earth and the need to protect it. Originally scheduled for April 22, Earth Day activities started earlier than that and continued throughout the year. In March, a four-day "teach-in for survival" at the University of Michigan involved more than fifty thousand people. "The Environment has just been discovered by the people who live in it," Barry Commoner wrote in 1971. "It was a sudden noisy awakening," he noted. "School children cleared up rubbish; college students organized huge demonstrations; determined citizens recaptured the streets from the automobile, at least for a day. Everyone seemed to be aroused to environmental danger and eager to do something about it."[17]

LEGISLATION FOR POLLUTION CONTROL

The strength and intensity of the public reaction to the pollution of the Great Lakes and other environmental problems were remarkable. In response to a public clamouring for protection of the lakes and the surrounding land, legislative authorities scrambled to put legislation in place to control pollution and help conserve the environment. While some government pollution control and conservation efforts had begun in the late 1960s, the early 1970s saw a prolific growth of government environmental programs. The federal governments in Canada and the United States both established new environmental agencies in 1970.

President Richard Nixon used an executive action to form the Environmental Protection Agency, or EPA. The EPA's creation involved the consolidation of pollution control activities that had previously been spread out through several departments. Announcing the formation of EPA, Nixon declared that the time had come to "perceive the environment as a single interrelated system."[18]

In Canada, Prime Minister Pierre Trudeau's administration established a federal agency responsible for the environment, under the direction of the newly appointed Minister of the Environment. The new ministry, Environment Canada, had the authority to deal with pollution problems and, like its counterpart in the United States, took over the responsibilities of an assortment of other departments that had previously held responsibility for environmental matters.

In both countries the enactment of new pollution control legislation occurred with heartening vigour. In 1969 the U.S. Congress passed the National Environmental Protection Act, aimed at providing "a national policy which will encourage productive and enjoyable harmony between man and his environment."[19] The government passed the U.S. Clean Water Act and introduced new air-pollution legislation. A new Coastal Zone Management Act encouraged states to develop and implement management programs for the use of land and water resources in coastal areas, including the Great Lakes.

Meanwhile, in 1970, the Canadian Parliament enacted The Canada Water Act, designed to give the federal government the foundation for a strong role in management of interprovincial and international waters. In the following year the Province of Ontario passed the Environmental Protection Act, which forbade the discharge of damaging contaminants above specified limits. In 1972 the province also reorganized existing agencies to establish the Ministry of Natural Resources, with a mandate including the management and control of natural resource use in Ontario.

The two countries took a similar response to the environmental awakening of the 1960s and early 1970s. They aimed their legislation at the worst and most obvious environmental abuses and tended to deal with environmental problems on an issue by issue basis: air pollution legislation to reduce discharges from smokestacks, water pollution legislation to reduce point-source liquid effluent from the end of industrial or municipal discharge pipes, and various land-use controls. The list of laws developed during this period was impressive, but something was missing: the idea that all the elements and activities of the environment formed a whole; and thus the need to integrate and co-ordinate all the various laws passed to protect the environment. In a statement that still holds true, Richard Bilder remarked at the time:

> Law governing Great Lakes pollution continues to be a complex hodge-podge of proliferating and occasionally inconsistent laws, regulations and ordinances issued separately by the two federal governments and their various agencies, the eight riparian states of the United States, the Province of Ontario, and the hundreds of cities, towns and other local jurisdictions that exercise relevant authority.[20]

The failure to recognize the inextricable interconnections of land, water, air – and people – would prove to be persistent during the subsequent years of governmental efforts to control environmental problems in the Great Lakes region. One aspect that the governments did recognize, however, was that pollutants do not respect political boundaries, and that it was imperative to work jointly on the pollution problems of the Great Lakes environment.

THE 1972 GREAT LAKES WATER QUALITY AGREEMENT

Shortly after the IJC submitted its final report on pollution of the lower lakes in 1970, negotiations began for an international agreement on Great Lakes water quality. The IJC report had clearly documented the continuing pollution of the lakes and recommended the immediate adoption of remedial measures. It also recommended that the governments establish an international board to co-ordinate pollution control programs and to monitor their progress.

In June 1970, high level officials began to discuss the recommendations and the possibility of an international agreement to solve the pollution problems. Finally, in April 1972, Nixon and Trudeau ushered in a new era of co-operation on the water quality problems of the Great Lakes by signing the Great Lakes Water Quality Agreement.

The governments called this historic agreement a statement of their determination "to restore and enhance water quality in the Great Lakes System."[21] The agreement included specific objectives for eight water quality factors: microbiology, dissolved oxygen, total dissolved solids, taste and odour, pH, iron, phosphorus, and radioactivity. Perhaps most important, it contained a schedule for the reduction of phosphorus loadings into the lakes. The agreement specified that all large sewage-treatment plants within the Great Lakes basin had to reduce their phosphorus discharges to a level not to exceed one milligram per litre of effluent. The agreement also granted greatly expanded powers and responsibilities to the International Joint Commission. It added a new responsibility to oversee the governments' progress in meeting the water quality objectives, along with powers of revision or addition to those objectives. To carry out the revised mandate the accord established two new advisory boards: the Research Advisory Board and the Water Quality Board.

The agreement also identified issues requiring further research and empowered the IJC to investigate them. The first of these concerns was the water quality problems of the upper lakes (Superior and Huron). Additionally, the agreement required the IJC to study pollution of the lakes from "non-point sources," that is, pollution that does not originate from a pipe or smokestack.

Finally, the agreement directed the IJC to report annually to both governments on the progress towards meeting the agreement objectives. Like a doctor continually checking the health of a patient, the IJC had the responsibility for monitoring the condition of the Great Lakes ecosystem. In a subtle statement of optimism, the governments agreed to review the effectiveness of the agreement five years from its date of signing. They were sure that the pollution problems that had so badly harmed the Great Lakes could be brought under control.

In statements made on the signing of the Great Lakes Water Quality Agreement, both Trudeau and Nixon captured the sense of optimism that dominated the period. Trudeau, characteristically philosophical, remarked in a press release:

> This Agreement deals with the most vital of all issues – the process of life itself. And in doing so it contributes to the well-being of millions of North Americans for it promises to restore to a wholesome condition an immense area which, through greed and indifference, has been permitted to deteriorate disgracefully.[22]

Nixon emphasized the beginning of a new relationship with the lakes:

> In recent years, as we know, the quality of the Great Lakes' waters has been declining with ominous implications for 30,000,000 Americans and 7,000,000 Canadians who live near their shores. The signing today of the Great Lakes Water Quality Agreement represents a significant step forward to reversing that decline.[23]

PHOSPHORUS WARS

The major emphasis of the Great Lakes Water Quality Agreement was the reduction of phosphorus pollution, and government activity reflected this concern. Eutrophication of Lake Erie and the other lakes had become the symbol of human abuses of the environment.

The U.S. Secretary of the Interior Stewart Udall remarked, "Lake Erie is the big challenge." He called the lake "the best test case we have" and concluded, "If we can lick water pollution here in the next few years then we can lick it in the country at large."[24] The battle against phosphorus became synonymous with the battle against pollution and environmental destruction.

Municipal sewage was the major source of phosphorus in the Great Lakes. Although most of the phosphorus in sewage came from detergents and human excreta, other sources added to the problem, especially industrial wastes and agricultural runoff, which included fertilizers and animal waste.

The IJC recognized these varied sources and proposed specific measures to deal with each. It recommended that an integrated program of phosphorus control include, first, the immediate reduction "of phosphorous content of detergents and the total quantities of phosphorous-based detergents discharged into the Great Lakes system." Second – but still urgent – was the reduction of the rest of the phosphorus that came from municipal and industrial waste effluents, "with a view to achieving at least an 80% reduction by 1975 and thereafter additional reduction to the maximum extent possible by an economically feasible process." The third line of attack would be to reduce the phosphorus discharged into river and lake waters from agricultural activities.[25] The IJC proposal aimed at dealing with all the major sources of phosphorus to the lakes, but what resulted was a government approach that focused largely on only one of the control options readily available: construction of sewage-treatment plants capable of removing phosphorus. For government, this path represented a way of doing something about the problem that did not fundamentally challenge established business interests and patterns of behaviour.

Both federal governments encouraged construction of sewage-treatment plants capable of removing phosphorus from municipal waste-water. The EPA and Environment Canada duly gave out massive financial grants for construction of these facilities to state, provincial, and municipal authorities. But for this kind of phosphorus reduction strategy to work effectively, the municipal waste-water systems had to collect all the sewage reaching the lakes, and, unfortunately, not all people and places in the Great Lakes basin were connected to a sewer system. In addition, many cities had combined their municipal sewers with storm sewers and in periods of heavy rainfall untreated wastes simply washed into the lakes. Despite these limitations, the construction of sewage-treatment plants remained a number-one governmental priority.

Canada had initiated a program for the collection and treatment of municipal wastes much earlier, before the IJC had begun its inquiry into pollution of the lower lakes in 1964. Capital expenditures in Ontario on all municipal sewage treatment works in the Great Lakes Basin, during the 1965-72 period inclusive, amounted to $1,284 million. One year after the signing of the Water Quality Agreement, sixteen new municipal treatment plants were operating in Ontario and extensive improvements were completed at eighteen others.

U.S. communities contributed a larger proportion of the Great Lakes phosphorus pollution but the construction of sewage-treatment plants proceeded more slowly there than in Canada. The signing of the agreement, however, did accelerate the municipal phosphorus removal program. In 1973 the IJC reported, "Since the signing of the Agreement on April 14, 1972, Federal grants totalling $403.4 million have been approved for 107 Great Lakes projects."[26] The total cost of these projects, including state and municipal costs, was estimated at $627 million.

Construction of sewage-treatment plants was, nonetheless, not the only available approach to reduction of phosphorus entering the lakes. A logical and efficient way to deal with a significant portion of the problem was the elimination or reduction of products containing phosphorus – something that a public becoming more and more conscious of environmental questions was now demanding. Soap industry officials countered, however, by arguing that the removal of phosphates at the later sewage-treatment stage would be not only less costly but also the only way they could ensure a "clean wash." The Canadian detergent industry put its public relations officers into full gear, sending out press releases warning that if phosphate bans were implemented, the move "would be equivalent to setting back cleanliness standards more than 20 years." Environmental control would have dire consequences for women working in the home, who "would find it virtually impossible to continue to enjoy the benefits of the modern automatic clothes washing machine." If this did not strike fear into enough hearts, the press release also argued that "any legislation which requires that phosphates should be removed from detergents would be unrealistic and scientifically unsound."[27] What industry managers really meant by this was that they feared the impact of a phosphorus limitation on their profitability.

Despite the objections of industry, the Canadian government instituted a 2.2 per cent phosphorus limit by weight, effective at the end of 1972. The U.S. government proposed similar legislation for enactment but could not overcome the stiff industry resistance and the legislation eventually went down to defeat. The industry claimed that it could not find a safe substitute for phosphorus – despite the fact that soap manufacturers in Sweden had been using the chemical NTA as a substitute with no discernible adverse effects.

Acting under pressure from student and consumer groups, the City of Chicago was the first of several Great Lakes cities to defy the companies by instituting a phosphate ban. Other cities followed suit, and state governments eventually adopted a limitation on phosphorus in detergents. Ohio, one of two states (the other was Pennsylvania) to ignore demands for a phosphate ban, finally imposed one in 1988.

Governments also implemented controls on some industrial sources of phosphorus. Within a year of the signing of the Great Lakes Water Quality Agreement, the International Joint Commission reported that both countries were passing extensive new antipollution legislation and taking administrative action. The battle to control eutrophication was on.

The fight to revive the lakes had been motivated by a public sickened and shocked by the damage that had already been done. At an IJC hearing in Thunder Bay, Ontario, one local resident, M. Bates, articulated some of the concern. He told the commission, "It too often appears to me that the question being asked is usually in the nature of 'how *much* can we pollute this or that body of water?' When the question should more properly be 'how *little* can we pollute it?'"[28]

Encouraged by a public clamouring for fishable, swimmable, and drinkable water, governments had begun taking the pollution problems of the lakes seriously. This was especially the case when they could do so without threatening corporate profitability or challenging social behaviour. Nonetheless, they were tackling the visible sources of pollution and environmental damage, and there seemed to be a real potential for improvement in the quality of the lakes. Government officials expressed strong hope that the money and effort being devoted to pollution control would truly reverse the environmental devastation of previous decades.

Only a short while after the signing of the 1972 agreement, the commission reported that "further degradation of the water quality of the Great Lakes may have already been slowed down."[29] Two years later the IJC was again able to say that all concerned were making progress. By 1974, however, the progress was being slowed. The IJC chastised the United States for failing to complete the necessary measures to reduce the phosphorus problems significantly. By the end of 1974 adequate sewage-treatment facilities existed for only 46 per cent of the U.S. population connected to sewers, while Canada was providing phosphate removal for 85 per cent of its sewered population. The IJC noted that about a quarter of the population of the Great Lakes Basin had no sewer connection to municipal facilities, and that "The extent of their pollution load to the lakes (including phosphorus) is unknown."[30]

The fourth annual report of the IJC identified the increasing worries about combined sewer overflows. The commission said the "programs to control pollution from the overflow of combined stormwater and sanitary sewers" were "fragmentary and obviously inadequate."[31] This problem would continue to be an obstacle to the expensive and extensive efforts to control the input of phosphorus to the lakes. Similarly, the governments did little throughout the 1970s to tackle the input of phosphorus from agricultural runoff.

Throughout the 1970s the countries continued to build new sewage-treatment plants and to expand the capacity of and make other improvements in many existing facilities. The progress was slow and did not live up to the expectations of the Great Lakes Water Quality Agreement. Nonetheless, there were positive developments. In a headline that contrasted starkly with the presentations of newspapers and magazines in the 1960s, the *Cincinnati Inquirer* headlined a story on the Great Lakes as "The Great Lakes: Born Again."[32]

The problems of pollution and destruction of the environment were far from over in the Great Lakes region. But by the mid-1970s citizen and government attention had slowed – and in some cases halted – water and air pollution and some of the other abuses of the previous decades. Other people were making new efforts as well to revive the beleaguered freshwater fisheries.

SEVEN

The Struggle to Revive the Fishery

BY THE EARLY 1950s the sea lamprey had thoroughly colonized Lake Superior.[1] As the eel-like fish had gradually moved up through the chain of Great Lakes, the destruction of lake trout and whitefish had been devastating. Now, to dull its attack in this last and largest lake seemed an almost impossible task.

Efforts to control lampreys had begun on Lake Huron in the early 1940s. Serious anglers from Rogers City in northern Michigan had suggested that the fish could be caught at small dams, or weirs, on their upstream spawning runs. During the Second World War this group convinced the Michigan conservation department to set up a weir on the Ocqueoc River, and also secured the joint support of the State of Michigan and the U.S. Fish and Wildlife Service for a program to have state biologists study the lamprey. The state went on to construct more weirs but none of them proved able to completely block the movement of lampreys into the spawning rivers.

Each successive effort, nonetheless, did provide valuable information on the life cycle and habits of the fish. The scientists learned that the adult lampreys migrated in spring and early summer into the tributary streams to spawn on gravel beds; and that their life cycle ended at the completion of only one spawning run. After hatching, the larvae migrated from the nest and burrowed in sand and silt areas of the stream, where they remained for three to seventeen years or in some cases longer. Undergoing their metamorphosis to an adult in the stream bed, the lampreys eventually emerged and began their downstream journey to the Great Lakes.

By 1947, improvements in construction and design had led to the operation of the first lamprey-proof weir on Carp Creek, twenty miles from Rogers City. It was found that if weirs and traps were properly constructed, they could capture entire spawning runs. Traps, in some cases, could also take in the young lampreys migrating downstream. The weirs, however, needed constant maintenance. Improper operation or destruction of a weir for even one day during the lamprey spawning season would allow the fish to repopulate the stream. Floods, ice, and logs were all potential hazards to successful operation of the weirs. In addition, weir construction and maintenance were costly, providing a further deterrent to their construction on all potential spawning rivers.

Over the next few years fishery officials of the U.S. Fish and Wildlife Service tried a whole host of methods for controlling the lamprey: everything from electrical shocking devices and sound and lights to fine mesh screens designed to strain the stream flow and catch the newly hatched adults before they entered the lakes. These efforts – which can best be described as technological quick fixes – were guided by an overriding belief in human engineering and technological mastery over nature. They succeeded best as illustrations of human insensitivity to the complex causes and subtleties of environmental problems.

The officials and scientists involved proposed other more bizarre and potentially dangerous control methods. One was the mass poisoning of spawning populations, fortunately rejected because it would also have killed off most other life in the water. Another proposal was to bulldoze the stream bed up onto the banks, thereby destroying the lamprey habitat. "With relief we find that this suggestion was rejected," noted Tom Kuchenberg somewhat ironically, "possibly because the destructive potential of the cure approached [that of] the disease."[2] Instead researchers began investigating the possibility of using a poison selective only for the lamprey. By 1951 this effort was in full swing at Hammond Bay, Michigan, where an abandoned coast-guard station had been converted into a lamprey-control research centre. Lampreys were indeed a deadly threat to the fisheries of the Great Lakes, but they also had one redeeming quality: in a way never before possible they brought together in one consulting group the fisheries management agencies of the Great Lake states, the province of Ontario, and the U.S. and Canadian federal governments. For decades, the fisheries agencies had held meetings and conferences to better co-ordinate their programs. None of these efforts had been very successful. Lampreys, however, proved to be a highly effective catalyst. After years of frustrated negotiations, Canada and the United States signed the Great Lakes Fishery Convention in 1955, resulting in the formation of the Great Lakes Fishery Commission. Its major focus was to co-ordinate research and management of Great Lakes fisheries; this included finding a solution to the lamprey problem.

The commission immediately began co-ordinating the lamprey-control efforts. Soon afterward, the Hammond Bay research station announced a preliminary laboratory success in killing lampreys with one of the many chemicals tested. The laboratory director, John Howell, walked into the lab, and found "larval lampreys dead and the trout still alive and happy" in one of the test jars containing the almost unknown chemical 3-Bromo-4-nitrophenol.[3]

The laboratory had tested over six thousand chemicals but this was the only combination found that would kill only lampreys. The chemical was so obscure that scientists around the continent had less than two ounces of it on their laboratory shelves. Now, scientists and chemical companies together worked to produce more of it for further research and possible manufacture. But despite the initial enthusiasm for the discovery, questions remained. Could this chemical "lampricide" be manufactured economically and with sufficient purity? Could it be released slowly in a controlled manner? Would it actually work in a stream? Would it have any unforeseen effects?

To begin answering these questions, researchers conducted testing in a simulated stream built at the research facility. The results seemed positive, so they made preparations to conduct the first out-of-doors experiment, to be held in October 1957. Dow Chemical produced a batch of the chemical for the experiment and it was released at a concentration of thirty parts per million at Little Billie's Creek in Cheboygan County, Michigan. The test killed 96 per cent of the lamprey larvae and had little effect on other fish.[4] It was the first time scientists had come up with a vertebrate toxicant that was selective for only one organism.

Despite the apparent success of the chemical tests, time was running out. There were only a few small surviving stocks of lake trout. So fishery officials decided to supplement existing populations with hatchery fish. Fortunately, Russ Robertson, a fish hatchery manager at Marquette, Michigan, had started gathering lake trout eggs from various parts of Lake Superior in the 1940s. Because lake trout are generally not prone to lamprey attack until their third year, the fish could be stocked in the lake before the lampricide had been applied to all the lamprey spawning streams.

Within three years scientists had treated all fifty-five lamprey infested streams in Lake Superior with the larvicide. By 1962 – four years after the stream treatments began – the lamprey population had plummeted to less than 20 per cent of its former abundance. To ensure continued control of lamprey spawning, the streams were treated again four years later. By 1966 only 3,475 lampreys were captured on their upstream spawning runs – compared to the almost 70,000 captured in 1961.

With this initial success the lamprey-control program and trout restocking shifted to Lake Michigan in 1960 and to Lake Huron later in the decade. Tom Kuchenberg summarized the battle against the lamprey: "The achievement of the larvicide is a truly impressive piece of work.... Foresight by Russ Robertson provided stocks, and swift, coordinated work by several governments and agencies helped to preserve remaining native trout in Lake Superior."[5] Questions remained, however, about the long-term consequences of using the larvicide. The lamprey was not completely eliminated from the lakes and concerns were raised about the eventual adaptability of the fish to the chemical poison and about the difficulties of using the poison in some rivers.

By the end of the 1960s, however, the unhindered ravages of the lamprey had been checked, at least temporarily, and the last remaining members of the native lake trout population were saved from what had appeared to be certain extinction. The protection of the lake trout population of Lake Superior and the restocking of lakes Michigan and Huron helped to undo some of the damage done to the fisheries of the Great Lakes, which had started with overfishing and was compounded by pollution and the purposeful introduction and accidental entry of exotic fish into the lakes. The fishery ecology was by no means secure, but at least one more successful effort had been made to stem the tide of decay.

THE ALEWIFE: A BLESSING IN DISGUISE?

In the mid-1960s the other marine invader, the alewife, began to die in almost unbelievable numbers. Their rotting bodies littered the waters and shoreline of the Great Lakes.

The most extensive dieoff occurred in the summer of 1967 in Lake Michigan. The incredible increase in the alewife population had peaked there in the fall of 1966 and that winter the dead fish started to appear. The deaths continued until past the midsummer of 1967, and involved several hundred million pounds of fish. "A dieoff of such magnitude had never been experienced in the Great Lakes," wrote scientist M.R. Greenwood, the author of a U.S. federal government report on the alewife dieoff. His report stated:

> Tons of dead and decaying alewives washed onto beaches and presented property owners and municipalities with expensive and nearly hopeless cleanup problems. The resultant loss to industry, municipalities, and recreational interests was estimated in excess of $100,000,000.[6]

Though the exact causes of the alewife dieoffs are not known, one theory is that their numbers simply exceeded the capacity of the environment to support them. Another is that the alewives are sensitive to the stress of low temperatures.

As municipalities and beach property owners were busy cleaning dead alewives from their shores, sometimes with bulldozers, fishery biologists were wondering aloud if the alewife problem might be turned to advantage. The alewife, they suggested, could provide a food source for highly valued predators such as the lake trout. An article in *Michigan Conservation* magazine in 1963 explained the argument:

> The alewife as a matter of fact may be a blessing in disguise.... it is known that the lake trout will eat large numbers of alewife, so that if the sea lamprey can be controlled and the lake trout re-established, the alewife will provide a ready form of food for these big, delicious, and

> highly prized trout. One estimate indicates that if the lake trout return to their former abundance in the Great Lakes, they could consume as much as 90,000,000 pounds of alewife each year.[7]

The lamprey invasion, overfishing, and the introduction of the alewife and other exotic species had decimated the Great Lakes commercial fishing industry during the previous two decades. The remaining commercial fishers looked with hope upon the success of the lamprey-control program and the restocking efforts for trout as the salvation of their dying industry. A revived trout population, they felt, would reduce alewife abundance and restore at least partially ecological stability to the lakes. But, at the same time, competing demands for fish resources were developing.

INTRODUCING THE SALMON

Beginning in the early 1960s North America experienced a recreation boom – a boom at least in part attributable to the increasing isolation and alienation of humans from the environment. Human economic activities and structures dominated the Great Lakes region of the 1960s. Over major parts of the region the natural environment had been reduced to rows of neatly ploughed fields and a proliferation of concrete, steel, and asphalt. The population of the region, which had become increasingly urban, became further removed from the natural world, except for quick summer or weekend trips to parks, lakes, cottages, and beaches. The pressures and pollution of the cities, traffic-congested streets, and summer heat all encouraged an overwhelming desire to "get away from it all." And get away from it they did.

One of the most popular outdoor recreational activities was fishing, and the Great Lakes formed a powerful magnet for people escaping the rush of the crowded urban centres. The recreational boom happened just when the largest and most prized Great Lakes fish had been severely depleted by the marauding lamprey. Sports fishers need fish, the larger and more aggressive the better. To sustain the interest in recreational fishing, they had to have hopes of catching something larger than a six-inch alewife.

With the recreational economy of many Great Lakes communities demanding that the interests of anglers be served, Great Lake fisheries managers began to look into the problem of ensuring a lively supply of sport fish. Officials noted especially the great masses of available alewives – a prey in need of a predator. They thought that restoring lake trout might be one answer. Then one agency took the leap. The Michigan Department of Natural Resources announced that coho salmon would be released into Lake Michigan streams by the state's Department of Conservation in the spring of 1966 and again in the following two years. This plan raised one special problem: coho salmon had never been permanently established anywhere in the world outside their traditional salt-water environment on the northern Pacific coast. Establishing them in the fresh-water environment of the Great Lakes was, to say the least, a challenge. Nobody knew what impact the salmon would have on native fish. Once again the move represented a technological quick fix to address a complex environmental problem.

In announcing the release program, the Michigan Department of Natural Resources explained the rationale for its decision:

> The coho is aimed at a specific fisheries management problem – namely to elevate the fisheries resource of the Great Lakes to its maximum potential for recreational fishing. The challenge in adopting the coho to the freshwater environment of the Great Lakes is an intriguing one.[8]

An intriguing challenge, indeed: the "ultimate aim" of the project was essentially to convert the upper Great Lakes' estimated two hundred billion pounds yearly of fish – mostly alewives – into "an abundance of sport fishes."

When it announced the salmon stocking program, the Michigan Department of Natural Resources also noted that it wished "to restore the depressed commercial fisheries to a productive and economically viable industry" – a goal clearly of secondary importance.[9] In fact, it became quickly apparent that in the eyes of state management the interests of commercial fishers were not only less important than those of sport fishers, but they were also in direct conflict with them. Because of the decline in fish stocks, commercial fishers had intensified their fishing efforts, developing an intensive gill-net fishery that utilized millions of feet of inexpensive nylon nets. Recreational fishers were not overly pleased to see these long strings of gill nets in their lakes – especially when their salmon and trout were likely to be caught up in them. The architect of the stocking program recognized this conflict, and made his choice:

> Our management scheme to eliminate the sea lamprey, restore predator fish, and control the alewives while providing a new Great Lakes sport fishery, collided with the gill-net fishery, since lake trout and salmon are highly vulnerable to gill nets. In our rehabilitation program, we planned stringent controls on gill nets, which are synonymous with commercial fishing.[10]

So the move to revive and protect recreational fishing restricted the catch of the commercial fishers. While limits on gill netting were undoubtedly needed, the message of the policy was clear: commercial fishers were not to interfere with the prized fish of the sports fishers. The state placed limits on the number of commercial licenses available, on the type of fish the industry could catch (the state banned coho and perch from the commercial catch in 1970, for example), and on locations for fishing. In short, the Michigan policy contributed to the further demise of the Great Lakes commercial fishers. In Michigan their number dropped from six hundred to two hundred.

Sports fishers responded with an intense enthusiasm for their new trophies. In *Michigan Conservation* journalist Russell McKee painted an unpretty picture of the new fishing fever:

> What could be so unusual as to drive a tight-fisted penny pincher to a cheerful squandering of $2,000 of hard-earned cash? What would cause dozens of peace-loving citizens to assault each other bodily with fists and poles to the extent of considerable personal injury? ... What would cause hundreds of men to go away on a treacherous sea – against all advice – in rubber rafts, bicycle boats, floating cars and trucks, canoes, and a host of other illogical craft?[11]

McKee's answer to these questions, of course, was "simply the prospect many men have of catching many big fish," namely the newly and artificially reared coho salmon. The salmon releases had a substantial impact on the economy of the Great Lakes region. By 1977 fishers were spending between ten and fifteen million angler days per year off the Michigan shore, compared with seventeen million for the entire west coast of the United States. Boat registrations in Michigan increased from 399,000 in 1965 to 487,000 in 1977, and the state's recreational boating fleet became the largest in the country. The sport-fishing industry in Michigan was contributing $250 million a year to the Michigan economy.

The fever for Pacific salmon was not restricted to Michigan waters. Lake Ontario had lost its native salmon population over eighty years earlier, leaving few large predators to feed on the abundant alewives and smelt. In 1969 the Ontario Department of Lands and Forests released 130,000 coho smolt, young salmon, in Lake Ontario, a move that was an instant success with anglers. John Power, an outdoor writer for the *Toronto Star,* described the salmon-stocking program as "a genie that couldn't be stuffed back into the bottle."

> Those tiny coho smolt returned in 1970 as huge silver-plated acrobats that propelled lines and pumped adrenalin. A sport fishery seemed to have been created overnight, answering the wishes of thousands of recreation starved southern Ontario anglers.[12]

The success of the coho salmon prompted governments to release other large predators, including the chinook salmon, which grows to an even greater size than the coho. By 1983 agencies had planted over 108 million chinook in the Great Lakes. Within the space of a few short years the salmon release programs had successfully restored the Great Lakes to prominence as an attractive, resource-rich haven for sports enthusiasts; and by the end of the decade the estimated economic value of the Great Lakes sports fishery was $4 billion. The planting of salmon, said Jim Wood, a staff member with the Michigan United Conservation Clubs, "was a message that we were not going to sit back and watch our lakes die."[13]

The introduction of the salmon did nothing to revive the sagging commercial fishing industry. Many of the remaining commercial fishers seriously questioned the salmon stocking program as the best way to revive the fishery. As the editor of the journal of the Ontario commercial fishery put it in May 1981:

> If the money that has been spent on planting chinook, coho and a variety of other species had been spent on pollution control and programs to combat the lamprey, alewives and smelt, we would be a lot further ahead than we presently are. Why support an expensive put

> and take industry with tax dollars when a naturally reproducing population inhabiting clean waters is within reach?[14]

As for the long-term ecological implications of introducing salmon into the Great Lakes, with the exception of those in Lake Superior the Pacific salmon have for the most part failed to reproduce naturally. According to one estimate only 9 per cent of the salmon in Lake Michigan are the result of natural reproduction.

The salmon-stocking program offered the promise of a quick, easy solution to the fisheries problem. This strategy, however, failed to restore ecological stability. The existing populations have to be maintained through yearly stocking of salmon, so that large-scale human manipulation of the fisheries has to continue. There are also uncertainties about the survival of the food supply necessary to maintain the prolific growth of the salmon.

Following the collapse of the alewife populations in 1966-67, their numbers continued to fluctuate dramatically, with no discernible pattern. The stocking of the lakes with salmon and to a lesser extent with trout increased the predator pressure on the alewife. The hatchery-released fish were soon eating up to 30 per cent of the alewife population. As ecologist James Kitchell noted, "That may not sound like much, but the alewife is a clupeid fish, a herring-like fish ... when levels of exploitation approach 40-50% of total production, virtually every clupeid stock in the marine environment has collapsed or just disappeared."[15] The salmon-restocking project depended not just on the short-term availability of alewives as food, but on the very survival of alewives as constant food source well into the future. The crash of the alewife populations would be disastrous for the survival of the salmon. This information, however, did little to dull government enthusiasm for salmon stocking. In 1989 Michigan DNR decided to stock about one million more than its usual 2.4 million chinook in response to dramatic declines in the Lake Michigan chinook catch. But as James Kitchell notes, "If chinook are dying – in part because of reduced alewife populations – adding more chinook is the wrong solution."[16]

The fishery remained unstable. Moreover, as excitement developed over the large size of the fish being caught, the problems of regional pollution had not gone away. New dangers to the fishery lurked on the horizon.

EIGHT

Toxic Chemicals:

The Great Invasion

IN 1968 officials of the U.S. Food and Drug Administration seized thirty-four thousand pounds of Lake Michigan coho salmon. It turned out that the fish, ready for sale to the public, had recorded levels of DDT almost double the permissible rate. A few years earlier, government scientists investigating herring gulls on Lake Michigan found that the birds' bodies contained more than two thousand parts per million of pesticide residue.

Both of these scientific teams had discovered a new threat to the health of the Great Lakes environment. The increasing concern about toxic chemicals was catapulted into headlines around the basin.

A German scientist had first produced DDT, short for dichloro-diphenyl-trichloro-ethane, in 1874. It was not until 1939, however, that its insecticidal properties were discovered by Swiss scientist Paul Mueller. During World War II, army doctors relied on it to fight the spread of lice and malaria-carrying mosquitos. After the war, scientists and industry alike hailed the chemical compound as a saviour in the battle against insect-spread diseases and an invaluable ally in every farmer's fight against insect pests. By 1962, DDT – one of the most powerful and long-lasting poisons ever concocted – was in everyday use. Environmentalist Rachel Carson, the author of *Silent*

Spring, a pioneering book on the hazards of pesticides, wrote, "DDT is now so universally used that in most minds the product takes on the harmless aspect of the familiar."[1]

DDT, however, was not harmless. Animal experiments showed that as little as three parts per million could inhibit essential enzyme activity in the heart, and only five parts per million could disintegrate liver cells. Most significantly, DDT, like other similar chemical compounds, had the ability to be passed from one organism to another and, in the process, to bioaccumulate – to concentrate to high levels.

Bioaccumulation occurs because many chemical contaminants taken in by animals end up being stored in their fat tissue rather than being excreted. When a fish eats another organism, a snail for instance, it takes in chemicals consumed by that snail. The fish in turn, when eaten by a larger fish, will pass along the chemical residue from all the creatures it has dined upon. Organisms at or near the top of the food chain, such as salmon, mink, fish-eating birds, and humans, can therefore accumulate large quantities of a chemical, even if their water source appears to be relatively safe. The chemical residues in gulls and Lake Michigan salmon were graphic evidence that organisms in the Great Lakes were vulnerable to this phenomenon.

In addition to accumulating chemicals taken into their bodies from their prey, fish also accumulate them through the gill during the respiratory process. Chemical contamination of fish through the gill is an especially significant source of exposure to persistent organic chemicals such as DDT, PCBS (polychlorinated biphenyl), and mirex. The solubility of all these chemicals is low in water and high in lipids, or fats. Fish gills consist of lipids, which absorb fat-soluble chemical compounds as well as oxygen from the water and pass them into the bloodstream. Through these processes of respiratory intake and bioaccumulation, the concentrations of DDT, PCBS, and other compounds found in the tissues of Great Lakes fish became anywhere from ten thousand to a million times greater than the levels found in the water.

THE FIRST TOXIC VICTIMS: BIRDS AND FISH

In the 1960s, with each passing year new evidence underscored the severity of the toxic chemical problem. Unrestricted dispersal of DDT had led to its accumulation in many of the world's ecosystems, including the Great Lakes. Although its use had diminished by 1970, it proved to be a persistent chemical that did not break down easily, and high levels accumulated readily in the lakes.

The majestic bald eagle, which had once flourished on the shores of the Great Lakes, was one of the early casualties. By the early 1970s, less than two dozen pairs of the eagles remained nesting along the entire length of the Great Lakes shores. The bald eagle's decline more than coincided with the introduction and use of DDT after World War II.

Bald eagles had long life spans and held a position as the terminal link in the food chain, making them especially susceptible to the bioaccumulative property of DDT. After accumulating large quantities of DDT in their bodies, their reproductive success began to slowly decline. Researchers studying the birds found that the DDT-contaminated eagles were laying eggs with abnormally thin or soft shells. Painstaking research over a period of more than two decades led to the discovery that DDT had gone one step further in the birds' bodies, breaking down through a metabolic process into a new compound called DDE. The scientists found that the more DDE contained in the female bald eagle's system, the thinner the shells she produced; and that during the period of incubation these shells would usually crack or be crushed. According to biologist Sergej Postupalsky:

> The extremely thin-shelled bald eagle eggs from Great Lakes aeries were found to be heavily contaminated with DDE, as well as other pollutants. Scientists were therefore not surprised with the near-zero breeding success of bald eagles nesting closest to the Great Lakes, the bodies of water containing the most DDT.[2]

But the bald eagle was not the only bird at risk. Other scientists studying the populations of gulls, terns, cormorants, and herons observed dramatic declines in the breeding success of these fish-eating birds. Michael Gilbertson, a young scientist with the Canadian Wildlife Service, travelled to a common tern colony located on two small islands in Hamilton Harbour in 1970:

> As I walked about one of these islands, many birds whirled around my head and swooped down upon me again and again to prevent me from approaching nests. Their piercing cries rang in my ears. As I wandered about I soon noticed something fundamentally wrong with the colony. While some young of varying age were found in the nests, the

> eggs in most had failed to hatch. On examining one of these eggs, I found that the young chick had died before it could crack open the shell. Several other eggs contained dead embryos.[3]

The real shock for Gilbertson came when he found a deformed two-week-old chick. The upper and lower parts of the bird's beak were twisted so they didn't meet, which meant certain starvation. As Gilbertson walked through the tern colony he wondered, "Was I witnessing part of a tragedy of major proportions?" Indeed, the phenomena observed in Hamilton Harbour proved to be painfully common throughout the Great Lakes region during the early 1970s.

In herring gull colonies on Lake Ontario in 1975 every ten pairs of gulls were successfully raising, on average, only 1.5 young birds. Normally, ten healthy pairs would have produced between twelve and fifteen youngsters. In the Lake Ontario colonies they were laying fewer eggs and many of those simply failed to hatch. The survival rate among the chicks that did hatch was dismal. On Lake Superior in 1975, 79.6 per cent of the herring gull eggs hatched, while on Lake Ontario, the more chemically-contaminated lake, the hatch rate was a pitiful 18.6 per cent.

Double crested cormorants were also in trouble. On Lake Ontario the population of this large, black, web-footed and fish-eating bird declined by 90 to 95 per cent, and in Georgian Bay by 50 per cent. Again, researchers were disturbed by the grotesque deformities: crossed bills, twisted skeletons, and skin and eye lesions. They believed that chemical contaminants were the cause of the deformities.

One scientist, Dr. Douglas Hallett, a chemist by training, used the newly-developed mass spectrometer to identify a multitude of frightening chemicals contained in the tissues of gulls. He reported that within a few short months his team "took the number of known contaminants from eight to fifty-seven or fifty-nine, somewhere around there." He noted grimly, "The gulls became the way we determined what chemicals were in the lakes."[4]

New analytical techniques and a heightened awareness of toxic chemicals prompted more discoveries of contaminants in the lakes and in the organisms they supported. Scientists found the pesticides chlordane and dieldrin among the chemicals in the tissue of gulls and other creatures. A chlorinated hydrocarbon like DDT, chlordane was liberally used in agriculture and spread on suburban lawns. Dieldrin, considered five to forty times as deadly as DDT, was also commonly used in agriculture – even after it had been found to cause hepatitis and fatal liver disease among workers in contact with the pesticide. Researchers also found mirex, hepatochlor, and, perhaps most worrisome, PCBS, one of the most pervasive and damaging chemical pollutants ever to reach the lakes.

Clear, colourless liquids, PCBS are amazingly stable chemicals that are not easily broken down even when exposed to temperatures of up to sixteen hundred degrees Fahrenheit. First produced commercially by Monsanto in 1930, PCBS have a resistance to fire, making them ideal for use in inks, detergents, plastics, hydraulic and heat transfer fluids, lubricants, and a variety of other products. The same characteristics beneficial to industry – stability and persistence – made them a deadly hazard to fish, birds, and animals in the Great Lakes environment. PCBS are easily bioaccumulated. Tests of herring gull lipids in the early 1970s found PCB concentrations of 3,530 parts per million, ten times the concentration of the next major pollutant.

Unfortunately, the contaminated fish, gulls, and eagles were not isolated instances of chemical problems in the lakes. Two years after the seizure of the Michigan fish, the Canadian government banned the sale and export of fish caught in Lake St. Clair. The chemical culprit in this case was mercury, not an organic pesticide but a heavy metal used in the pulp and paper industry. For over twenty years the Dow Chemical Company in Sarnia had dumped an average of 30 pounds of mercury a day – with occasional peaks of 250 pounds a day – into the St. Clair River. After it settled to the river bottom the elemental mercury was converted by micro-organisms into lethal methylmercury.

The Lake St. Clair commercial fisheries were substantial, providing about forty small family companies with between one million and two million dollars' worth of fish per year. The 1970 ban on the sale of Lake St. Clair fish caused the collapse of the fishery, financially devastating the fishers and their families. To compensate them, and to prove it did "mean business with regard to the policing of the environment," the Ontario government announced a $35,000,000 lawsuit against Dow in early 1971.[5]

Seven years later, after reaching an out-of-court settlement, Dow paid $100,000 to the province and $250,000 to the fishers: a paltry settlement indicating the reluctance of governments on both sides of the international border to tackle head-on the threat of toxic chemicals. Increasingly, toxic chemicals came to dominate the concerns of citizens, scientists, and others interested in the Great Lakes.

In 1973, unsafe levels of mercury were still present in fish from Lake St. Clair, and the contamination had spread to Lake Erie. The International Joint Commission reported:

> Although both countries have programs to control the discharge of mercury and heavy metals, concentrations above those recommended for safeguarding human health continue to be found in fish taken from the St. Clair River, Lake St. Clair, western Lake Erie and the International Section of the St. Lawrence River.[6]

The commercial sale of trout, salmon, and walleye from these waters was banned, and local fishers were the losers. Lake Erie commercial fisherman Don Misner recalled with bitterness, "We were made to suffer for it. It was a terrible thing, it was devastating."[7]

Researchers continued to find fish containing levels of toxic chemicals hazardous to human health in other areas of the lakes. They detected PCBS and DDT in salmonid species of fish in Lake Michigan and in Lake Huron's Saginaw Bay. The levels were high enough to necessitate a ban on their sale and prompt health agencies to warn the public about their consumption. In its 1975 report to the International Joint Commission, the Water Quality Board stated bluntly, "Toxic and hazardous materials represent a major threat to water quality and the fishery of the Great Lakes."[8]

As the 1970s unfolded, evidence of the toxic chemical contamination in the Great Lakes was revealed again and again. Mink farmers stopped feeding Great Lakes fish to their captive mink because of a reproductive failure apparently caused by accumulating pesticide residues from the fish. In 1975 researchers discovered that carp from Green Bay contained PCBS above the allowable limit of five parts per million, and the state closed down the four-and-a-half-million-pound fishery. That same year, the U.S. Food and Drug Administration seized one hundred thousand cans of Lake Michigan coho salmon because the levels of PCBS in the fish were between 7.6 and 10.9 parts per million, well over the 5 parts per million limit. In 1975 the International Joint Commission reported that fish in Lake Superior were accumulating increased levels of PCBS and mercury. Scientists also detected high concentrations of asbestos fibres in Lake Superior waters.

The list of chemical contaminants grew longer. In 1977 the International Joint Commission Water Quality Board noted that "thirty-eight previously undetected contaminants" had been found in samples of fish and wildlife from the lower lakes.[9]

THE CHEMICALS ARE EVERYWHERE

The onslaught of chemical production and use in the decades following the end of World War II increasingly harmed the environment, and by analysing sediment samples in the lakes, scientists have reconstructed the history of chemical contamination in the Great Lakes region. Two Environment Canada researchers studying the available information on chemicals such as PCBs, chlorinated benzenes, mirex, and chlorotoluenes found a "progressive increase in contamination from about 1915, with the highest levels occurring in the late 1950s through the early 1970s."[10]

Pesticide production in the United States expanded an astonishing 3,000 per cent between 1951 and 1977. Pesticide use in the United States increased from 34 million pounds in 1953 to 119 million pounds in 1965. More than 58 per cent of this went for agricultural use. The rest went into roadsides, golf courses, parks, hydro right-of-ways, forests, and suburban lawns and gardens. By 1964 the total market value of pesticides used in the United States was over one billion dollars and a few years later the U.S. Department of the Interior was conservatively estimating that thousands of pounds of pesticides were annually running off the land into rivers and lakes. The department warned that pesticide use was so loosely controlled that the world environment was "permeated with these substances."[11] By 1972 the U.S. side of the Great Lakes basin was using some forty-five million pounds of pesticides, more than one pound per person per year and including herbicides, insecticides, and fungicides. The Canadian side of the lakes, exclusive of Lake Superior, was using over thirteen million pounds (six million kilograms) of pesticides on field crops in 1973.

The lakes themselves had become the ultimate dumping ground for all the artificial substances applied to the land area of the basin. With the widespread pesticide use it was inevitable that chemicals such as chlordane, dieldrin, HCB, and HCBD would make their way into the lakes. But there was more: industries had also been directly discharging wastes into the lakes and river systems, adding yet another chemical burden to the Great Lakes environment.

In its 1975 report, the International Joint Commission's Water Quality Board identified over five hundred significant industrial facilities discharging pollutants into the Great Lakes and their tributaries. In St. Louis Bay, on western Lake Superior, researchers traced high phenol levels to pollution discharges from Conwed Corporation, U.S. Steel, and Potlatch Corporation. Similarly, cyanide, phenols, ammonia, and other compounds released by two Canadian Steel companies, Stelco and Dofasco, contributed to the destruction of the once valuable fishery and recreational potential of Hamilton Harbour on Lake Ontario. A consulting company report in 1973 denounced the water opposite the steel companies as being "unsuitable for any biological activity."[12]

Other industrial waste sources indirectly contributed to contamination. The Great Lakes states produced one-quarter of the hazardous waste in the United States and much of it was simply dumped into landfills, from where it slowly leaked into the environment. In March 1978, for example, scientists found very high concentrations of PCBS in fish collected in the Sheboygan River, a Lake Michigan tributary in northeastern Wisconsin. The levels of PCBS in the fish samples ranged from twenty-six to almost one thousand parts per million. The U.S. Food and Drug Administration (FDA) guideline for PCBS in fish was five parts per million, a level later lowered to two. The Wisconsin Department of Natural Resources (DNR), investigating the source of the contamination, quickly identified a metal-casting plant as the source. The plant had been disposing of PCB-contaminated waste along a dike that ran next to the Sheboygan River in an area behind the plant building. High water-levels, runoff, and erosion had deposited PCBS in river sediments, contaminating the food chain.

The problem was not confined to the United States. In Ontario a 1979 survey by the environment ministry found eight hundred previously unidentified waste dumps. The government admitted that this was only a fraction of the total number that existed. The problem of abandoned waste in Ontario was graphically illustrated in the summer of 1981 when a black gooey tar oozed out of the ground in a schoolyard in suburban Sarnia. Tests on the material revealed that it contained styrene, benzene, and polyethylbenzene. All of these chemicals were known causes of skin disease and suspected carcinogens.

As the magnitude of toxic chemical pollution emerged, it became obvious that control of damaging chemical pollutants was desperately needed. In response to vocal campaigns by environmental groups, governments had imposed restrictions on the use of DDT in both Canada and the United States in the early 1970s. As a result, scientists had seen a decline in the residues of this pesticide in fish by the middle of the decade. Ontario banned aldrin-dieldrin in 1969, and in 1974 the United States followed suit. By 1977 the state governments of Minnesota, Michigan, Indiana, and Wisconsin had banned the use of PCBS, and Canada stopped its use in new electrical capacitators and transformers. Alarmingly, the levels of the controlled substances did not drop as rapidly as expected in the tissue of fish and other organisms. DDT levels in Lake Michigan trout still exceeded the guideline of five parts per million in 1976. In the same lake, the levels of dieldrin remained above the U.S. FDA guidelines in both 1975 and 1976.

Despite control efforts the high levels of toxic chemicals persisted. The problem was that the easily identifiable point sources of pollution were not the only cases of toxic chemicals entering the lakes. The 1972 Great Lakes Water Quality Agreement had mandated the International Joint Commission to investigate the previously unexplored question of pollution from land use activities as well as the issue of water quality in the Upper Great Lakes. As these two studies progressed, it became evident that the water quality problems in the Great Lakes were not only more widespread and all encompassing than previously thought, but that the control of this pollution was going to be much more difficult than anticipated. More than anything else, the problem graphically illustrated the unseverable interconnections between the air, the land, and the water.

THE PLIGHT OF AN INTEGRATED ECOSYSTEM

In the early 1970s, Thomas Murphy, a chemist at DePaul University in Chicago, began analysing the chemical composition of rain falling on Lake Michigan. The lake receives about thirty-two inches of precipitation annually, about half the annual water input to the lake, and Murphy investigated the contribution of this rain to pollution problems. The results of his study were startling: he found that over 1,400 pounds of PCBS were washed into the lake each year by rain and snow. This amount of PCBS from the atmosphere represented about half the known inputs of PCBS to Lake Michigan.

Other studies of chemical contaminants in rain and snow confirmed Murphy's alarming findings: the atmosphere itself was a significant source of chemical contamination of the Great Lakes environment. Beginning in 1972 an International Joint Commission study team had examined the problem of pollution from sources other than direct discharge by pipes to the lakes. Their report, released in 1976, found that atmospheric deposition contributed toxic substances to the Great Lakes. Other International Joint Commission scientists studying pollution in Lake Huron and Lake Superior reached similar conclusions. Their report suggested that atmospheric inputs were responsible for up to 40 per cent of the loadings of certain pollutants to the lakes, including phosphorus, heavy metals (such as lead and mercury), toxic organic contaminants, and sulphur dioxide.

Human activities on land were also contributing to the toxic chemical problem in the lakes. The scientists studying pollution from land-use activities found that each year a staggering eleven-million tonnes of sediment washed from fields, roads, and forested lands into tributaries of the Great Lakes. Not only did this loss contribute to a major depletion of soil, but it carried with it large quantities of toxic chemicals and fertilizers attached to the soil. Pesticides and fertilizers bind easily to the small particles that wash off the land. On Lake Erie, for example, 896 tonnes of lead per year were contained in the sediment entering the tributaries to the lake, representing 40 per cent of the tributary load. Over one-half tonne of PCBS was also entering Lake Erie from its tributaries and 20 per cent of the PCBS in Lake Ontario were entering via tributary flow.

It was also becoming apparent, as political scientist John Carroll pointed out, that through agricultural runoff and other "nonpoint sources" the "nonindustrialized rural areas were contributing a greater pollution load to the lower lakes than had been previously suspected."[13]

As the evidence mounted it became clear that the pollution problems of the 1970s were not easily controllable. A large number of small contributors to pollution are much more difficult to control than one big source. No longer would relatively simple solutions, such as building sewage-treatment plants, suffice. The sources of contaminants were many and varied and widespread throughout the Great Lakes ecosystem. At the very least, the region needed innovative new management approaches that included co-operation and creativity.

More fundamentally, the prevailing capitalist ethos of expanding production and growth had to be challenged. Author Lee Davis summarized the problem:

> It is not that these entrepreneurs [the chemical manufacturers] are deliberately trying to poison us; the problem is that they all operate according to a system that compels them to produce as many goods as possible, as quickly as possible, at as high a profit as possible. This was as true of the very beginnings of the chemical industry as it is of the present.[14]

It appeared to many observers to be suicidal to perpetuate an economic and social system that encouraged the poisoning of the atmosphere, land, and water. They recognized the need for a new rethinking and reordering of the human-environment relationship in the Great Lakes region.

Most alarming about pollution and the toxic chemical problem was the apparently long time it was going to take the lakes to recover. Even if governments succeeded in stopping all the pollution inputs immediately, the large volumes of water in the lakes and the long water-retention times of toxic chemicals meant that it would take decades to clean up properly – if indeed the job could ever be finished. The continued high levels of DDT and dieldrin in the lakes, even after their use had been banned, were evidence enough of this. Dennis Konasewich, an International Joint Commission chemist, argued that it would take six years to remove 90 per cent of a pollutant from Lake Erie, even after all sources of the pollution had been expunged. The same process would take twenty years for Lake Ontario, one hundred years for Lake Michigan, and about five hundred years for Lake Superior.[15]

With these new insights, new approaches to Great Lakes problems clearly needed to be developed. The twisted beaks and reproductive failures of terns, gulls, and cormorants warned that the dangers of inaction could prove fatal.

THE 1978 AGREEMENT TO CONTROL POLLUTION

Prodded by a public increasingly concerned about the health of the environment, in the mid-1970s governments in Canada and the United States began legislative action aimed at controlling the problem of toxic chemicals. In Canada the federal government passed the Environmental Contaminants Act of 1976, which provided the basis for the regulation of the importation and manufacture of toxic substances. In 1977 the U.S. Congress made amendments to toxic chemical provisions of the U.S. Clean Water Act and passed a new act, the Toxic Substances Control Act. This Act provided the authority to regulate and control the manufacture of chemicals that posed unacceptable risks to the health of the environment.

Despite the appearance of wide-ranging powers to control chemicals, the legislative changes were limited in their scope. They dealt only with a limited number of chemicals and sources. Comprehensive and co-ordinated legislation was still needed, and even conventional pollution still posed problems for legislators. While improvements were made during the 1970s on phosphorus control and other more visible pollution problems, progress in many other cases was slow. Industries tended to view pollution control as a threat to their profits and resisted spending the necessary money for environmental improvements. On the other hand they were more than willing to spend money to ensure that their viewpoints were heard by legislators. Between 1977 and 1980 chemical-industry Political Action Committees alone spent two million dollars on campaign contributions to U.S. political figures. John Quarles, an Environmental Protection Agency administrator, concluded, "Private industry, driven by its own profit incentives to exploit and pollute our natural resources, uses its inherent advantages to exert political pressures to resist environmental requirements."[16]

In many cases the governments, sympathetic to the concerns of industry and fearful of their threats to move elsewhere, were reluctant to enforce controls. The programs they had put into effect were limited by a lack of funding, slow construction schedules, and general resistance. In 1975 the phosphorus loadings to Lake Erie were three times higher than the target adopted in the 1972 Agreement. "Municipal pollution abatement on the United States side has been hampered by the slow use of available funds," wrote the Water Quality Board in 1975.[17] Pollution control at Canadian pulp and paper mills on Lake Superior was described as "inadequate." If the region was going to make progress on the more difficult problem of diffuse sources of toxic chemicals, greater enforcement of controls and greater commitment to the issues would be needed.

Throughout the 1970s, the reports of the International Joint Commission and its boards emphasized the need for this stronger commitment. Beginning in 1973, the Water Quality Board had identified specific areas around the lakes where conditions had significantly degraded water quality. The 1977 Water Quality Board report listed forty-seven areas that did not meet the water quality objectives specified in the 1972 Agreement. Every lake had at least four "problem areas." These areas included the Toronto Harbour and waterfront, Hamilton Harbour, the Buffalo River, the Cleveland area, the Detroit River, Saginaw Bay, Indiana Harbor, Sault Ste. Marie and the St. Marys River, and the Duluth-Superior Harbour. The pollution problems of the lakes were both general and specific in their nature.

When the two governments signed the 1972 Great Lakes Water Quality Agreement they had agreed to revise it after five years. As this date approached, the two countries clearly needed renewed and strengthened commitments to reduce and reverse the damage done to the Great Lakes environment.

On November 22, 1978, U.S. Secretary of State Cyrus Vance and Canadian Secretary of State for External Affairs Don Jamieson signed a revised Great Lakes Water Quality Agreement. Jamieson hailed it as representing "the practical approach of our two peoples to resolving problems which affect us both, to which we are both contributors, but neither can solve alone."[18] A Canadian government press statement described the agreement as reaffirming "the determination of both countries to restore and enhance Great Lakes water quality."[19]

The new agreement maintained the basic structure of the 1972 Agreement, and added important improvements. "Additional programmes and measures to meet problems in the Great Lakes pollution which were not evident or fully understood at that time" were added, said the signatories.[20] In particular, the new agreement called for the following measures: more effective control of persistent toxic substances; identification of the quantities and types of airborne pollution entering the lakes; control of pollutants from land uses such as forestry and farming; and further reductions in the levels of phosphorus allowed into the lakes. It established a revised set of water quality objectives, including substantially more stringent standards for radioactivity and a number of toxic contaminants. The new agreement also contained a list of 350 chemicals that should be banned from the lakes and set target dates for municipal and industrial pollution abatement and control programs at the end of 1982 and 1983 respectively.

Additionally, the new agreement recognized the increasingly

obvious interconnectedness of land, lakes, air, plants, and animals in the region – to the extent of introducing the term "Great Lakes Ecosystem." The agreement defined ecosystem as "the interacting components of air, land, water and living organisms, including man, within the drainage basin of the St. Lawrence River at or upstream from the point at which this river becomes the international boundary between Canada and the United States."[21]

The change in wording brought hope that the Great Lakes environment would in future be managed in a way that recognized the connections between all the various components of the ecosystem. For centuries people had dealt with the problems of land use, fisheries, air pollution, and water pollution in isolation from one another. The wording of the new agreement indicated the possibility of new approaches to dealing with these problems.

The parties did agree "to make a maximum effort to develop programs, practices and technology necessary for a better understanding of the Great Lakes Basin ecosystem and to eliminate or reduce to the maximum extent possible the discharge of pollutants into the Great Lakes System."[22] At the same time, they congratulated themselves on their past efforts to control pollution and reverse environmental degradation. "Both countries have devoted great effort and substantial resources to the restoration and enhancement of water quality of the Great Lakes," said a Canadian external affairs press release distributed at the signing.[23]

While there could be argument about the extent of the efforts and the usefulness of the approaches of the two governments to control pollution, there was no doubt that some visible improvements in the quality of the environment and particularly Great Lakes waters had occurred since 1972. The ravages of the lamprey had been largely checked. Expenditures on construction of sewage-treatment plants, although less than anticipated, had resulted in the introduction of many new plants. The 1978 report of the Great Lakes Water Quality Board reflected the importance of these facilities:

> A substantial drop in point source municipal phosphorous loading to the Great Lakes basin has occurred. Aggregate Great Lakes concentrations of phosphorous (total load divided by total flow) in municipal point sources have dropped from 2.6 mg/L in 1975 to 1.8 mg/L in 1978.[24]

The board also reported "decreases in the concentration of DDT, DDE, dieldrin, HCB, mirex, and PCBS in herring gull eggs from Lakes Superior, Huron, Erie, and Ontario."[25]

There were a number of signs that, in certain ways, the lakes were returning to health. Local jurisdictions had been able to reopen some of the beaches closed during the early 1970s. Boaters were less likely to come across oil slicks and floating mats of algae. But for every step forward there seemed to be an equal number of steps backward. The 1978 Water Quality Board report had identified previously unreported compounds such as PCTS (polychlorinated terphenyls, which have environmental characteristics similar to those of PCBS) in herring gull eggs from Lake Erie. In addition, trace levels of dioxin were reported in fish from Lake Ontario and Saginaw Bay. Toxic chemicals continued to enter the environment in damaging concentrations, as the Water Quality Board emphasized: "The Board recognizes the importance and enormity of the task confronting the agencies involved in implementing laws to control toxic and hazardous substances."[26] Despite all the encouraging signs of improvement, the lakes and the surrounding land were still a long way off from receiving a clean bill of health.

NINE

The War Is Not Over

ALTHOUGH THE 1978 Great Lakes Water Quality Agreement reflected the concerns about toxic chemicals and land use that had emerged since the signing of the first agreement in 1972, new challenges had presented themselves, and old problems had proved to be ruthlessly persistent. In 1972, the job of controlling phosphorus had seemed massive and difficult. In 1978 the challenge – and the problem – had if anything grown larger still.

The 1978 agreement specified important new commitments in the battle against phosphorus pollutant problems but did little to increase the various governments' ability to meet pollution targets. As the International Joint Commission noted in its 1982 report, current estimated phosphorus loads for all the Great Lakes still exceeded the "proposed target loads" in the agreement.[1] Two years later, in 1984, the IJC again chastised the governments for their tardiness in upholding the commitments to reduce phosphorus pollution. Although the United States and Canada had spent $7.6 billion to construct and upgrade sewage-treatment plants, 39 of the 390 major facilities in the basin had missed the December 31, 1982, construction deadline. Other plants were having difficulties operating to design capacity.

The 1978 agreement called for sewage-treatment plants on the lower lakes to reduce their phosphorus discharge to a minimum of one milligram per litre. In November 1983 the Water Quality Board identified nine major sewage plants that had failed to meet this level, including facilities in Wyandotte, London, Toronto, Hamilton, Niagara Falls (N.Y.), Buffalo, Amherst, and two in Cleveland. Ineffective and improperly operating sewage-treatments plants continually undermined efforts to reduce phosphorus pollution effectively.

Many municipalities were also not dealing effectively with the longstanding problem of combined sewer overflows. A significant amount of phosphorus continued to reach the lakes via these overflow systems. The fact that many older communities, such as Toronto and Detroit, had sewer systems that were not originally divided into separate sanitary and storm sewers caused continuous, severe problems. In dry weather, matters went satisfactorily, with all the effluent from the sewers flowing to the sewage-treatment plants and being properly treated. But periods of heavy rain filled sewers to capacity, creating an effluent load too large for sewage-treatment plants to handle. The combination of sewage and storm water was therefore diverted – untreated or poorly treated – via overflow pipes to the lakes. A weekend of thunderstorms in July 1986, for example, forced health officials to close four Toronto beaches to swimming because of the high bacteria levels resulting in part from sewage overflow.

While obvious limitations and shortcomings remained, the major achievement of the 1972 and 1978 Great Lakes Water Quality Agreement was a reduction in the levels of point-source phosphorus pollution. The measures cut down the inputs of phosphorus from municipal sewage to only a fraction of what they had been during the early 1960s. This reduction led in turn to a noticeable improvement in the eutrophication problems that had plagued the lakes. Evidence of the improvement was soon apparent in the revived summertime crowding of Lake Erie beaches and confirmed by the people who relied on the lake for a living. In January 1988, staring out the window of his Port Dover fish processing plant, Don Misner declared, "The water quality now is great."[2]

The reduction of the eutrophication in the Great Lakes, and especially in Lake Erie, was so substantial that a 1985 book about Lake Erie was entitled *Erie: The Lake That Survived.*[3] One sign of improvement was the successful breeding of mayflies in the island area at the western end of the lake. Lake Ontario and Saginaw Bay, two other areas that had been hurt by eutrophication, experienced dramatic changes in the composition of algae species. Similarly, in Green Bay, biologist Paul Sager observed significant improvements in the algal species. Secchi disks, devices dropped into water to measure turbidity, could now be seen at greater depths than ever before in Green Bay waters.

Despite success in controlling nutrient pollution at the municipal point source, the governments were patently unsuccessful in their attempts to control phosphorus washing off the land and deriving from fertilizers, animal feedlots, and pet feces. The non-point sources of nutrients, if not checked, had the potential to undermine gains made in reducing phosphorus from municipal sewage. The IJC's PLUARG (Pollution From Land Use Reference Group) estimated that non-point inputs were responsible for 53 per cent of the phosphorus input in Lake Superior, 30 per cent in Lake Michigan, 50 per cent in Lake Huron, 28 per cent in Lake Ontario, and 48 per cent in Lake Erie. To address the eutrophication problems in the lakes, especially in nearshore areas, non-point source phosphorus inputs had to be reduced.

PLUARG suggested a number of specific programs for reducing non-point phosphorus loads to the lakes, including the planting of trees along stream banks; the preservation of wetlands that effectively act as filters for land pollution inputs; and education and financial assistance for farmers to assist them in adopting farming practices that reduce manure and fertilizer runoff. In addition to reducing phosphorus pollution, these measures would also cut down on soil erosion.

A 1983 IJC Task Force on non-point controls concluded, however, that although governments had followed some of the PLUARG recommendations, most of the conclusions were ignored. "In Canada there has been no action to develop a comprehensive program to address non-point sources of water pollution in the Great Lakes," said the report.[4] Its assessment of U.S. efforts was equally grim: "In the United States, there is no comprehensive program for the control of non-point sources of pollution."[5]

The failure to implement non-point phosphorus pollution control undermined the efforts to restore the lakes to a healthy condition. Reviewing the Great Lakes Water Quality Agreement, a Committee of the Royal Society of Canada / National Research Council remarked, "The lack of programs to address non-point-source control of phosphorous loading suggests a loss of 'will' to do more than attempt control of point sources of phosphorous pollution with rather old-fashioned methods."[6]

The failure to impose phosphorus limits in detergents in Pennsylvania and, until 1988, in Ohio, also helped prevent the achievement of the agreement's goals for phosphorus reduction to the lakes. These failures slowed and in some areas prevented the recovery of the lakes from the nutrient problems that had become so acute in the early 1960s.

In its 1985 report, the Royal Society of Canada / National Research Council committee concluded that conditions in the lakes with respect to nutrients, though still uncertain, had improved. It stated, "On the whole, water quality problems caused by enrichment or eutrophication have decreased since 1972." Saginaw Bay was singled out for showing marked improvement. There were still eutrophication problems in some nearshore areas of Lake Superior and Lake Michigan; and "Lake Erie remained relatively high in phosphorous content, possible because of a release of nutrients stored in sediments, but there are signs of a downward trend."[7]

The efforts to control phosphorus pollution had paid off, to a certain extent. Massive expenditures of money and twenty-five years of remedial work had stablilized or in some cases reduced the Great Lakes' phosphorus or enrichment problems. Yet if the gains of those years were to be maintained, the two nations would need still greater efforts to reduce the non-point phosphorus source and the number of combined sewer overflows. The 1987 report of the Great Lakes Water Quality Board concluded: "Difficulties in meeting the established target loads have been primarily attributed to excessive contributions from agricultural sources and non-compliance with the point source limit."[8] The mix of sewage, fertilizers, and manure had pushed these massive bodies of water to the brink of destruction, and it was by no means certain that this would not happen again. The need for replacement or repair of existing sewage-treatment plants in the future and the inevitability of an expanding population are just two of the factors that could jeopardize the qualified success of controlling phosphorus pollution. By the late 1980s the victory over eutrophication was neither complete nor final.

THE PERSISTENT TOXIC THREAT

In the minds of many onlookers in the region, the battle against phosphorus had become synonymous not only with the battle against pollution in the lakes but also with the overall environmental crisis in the region. Phosphorus, however, was only one of the environmental challenges to be faced in the Great Lakes region. In 1979, in an anniversary publication, *70 Years of Accomplishment*, the International Joint Commission identified the biggest problems confronting the Great Lakes region, and control of toxic chemicals, particularly those distributed by long-range transport, headed the list.[9] Eight years later, in November 1987, scientists from the Great Lakes Water Quality Board reported to a packed conference room in Toledo, Ohio, "The problem of persistent toxic substances has emerged as the major issue confronting the Great Lakes today."[10]

The 1978 agreement had emphasized the imperative of dealing with the toxic chemical problem, and the governments optimistically agreed to a goal of "zero discharge" of persistent toxic substances. As the next decade progressed, the issue of toxic and hazardous chemicals was clearly on the mind of the International Joint Commission, the public, and government officials.

The most damaging chemicals identified were the ones with essentially irreversible effects. The 1979 IJC report on the pollution of the Upper Great Lakes had recommended "banning the manufacture, import, and use of certain of these substances, including PCBS, PBBS, aldrin, dieldrin, DDT and its derivatives, and other persistent, highly toxic substances, whose entry into the environment is difficult to control, if their use is permitted."[11] The commission had also recommended that governments restrict the use of new chemicals until their safety was established.

To the credit of both countries, legislative initiatives provided the means to control chemicals more effectively. The governments placed restrictions or bans on the manufacture and use of chemicals such as mirex, PCBS, and PBBS. The United States introduced a Toxic Substances Control Act (TSCA) and made amendments to the Clean Water Act. The Canadian government passed the Environmental Contaminants Act, and the province of Ontario introduced new statutes and regulations aimed at greater control of contaminants.

Unfortunately, the approach taken dealt with chemicals on a case-by-case basis, which proved to be an almost impossible task. Although the new control efforts achieved almost immediate reductions in the levels of DDT, PCBS, and mercury in fish and herring gull eggs, by the early 1980s the downward trend in levels of these contaminants ceased and the concentrations remained constant or had increased slightly. For each chemical controlled, at least one other chemical – and sometimes more – presented an equally serious threat to the environment. The regulatory machinery simply could not keep up with the large number of chemical pollutants.

Any success was compromised too by a failure to deal with the varied sources of the chemicals. Many companies complied with limitations on their direct effluent discharges to the lakes by dumping wastes in landfills or by sending them through the sewers to sewage-treatment plants incapable of treating the wastes. The Water Quality Board's 1981 report to the IJC pointed out these shortcomings and suggested that there was an "absence of an overall Great Lakes ecosystem strategy for toxic substances control activities."[12] The comptroller general of the United States expressed a similar sentiment in his 1982 report: "Great Lakes pollution has yet to be comprehensively addressed."[13]

Toxic chemicals, it was found, were entering the environment in less direct ways. They came into lakes through slow but steady runoff from urban and rural areas. They were part of the sediments and pollutants that settled in harbours and other parts of the lakes, accumulating from years of discharge. They came from contaminated groundwater and fallout from the atmosphere.

Each year, water runoff from cities around the basin contributed large amounts of toxic chemicals to the lakes. A Canadian government study found the following substances entering the lakes each year from urban runoff: 420 tons of zinc, lead, copper, nickel, and chromium; 8 tons of cobalt, mercury, arsenic, selenium, and cadmium; close to a ton of PCBS; and substantial amounts of chlorinated benzenes and organochlorine pesticides. Other sources included lead from automobile exhaust (although new standards in both countries would reduce this amount), pesticide runoff, wind drift of sprayed pesticides in agricultural areas, and runoff from suburban lawns. U.S. householders applied between 2.4 and 4.8 kilograms of pesticides for each acre of suburban lawn, providing one of the most intense doses on the continent.

Past and present pollution discharges created highly contaminated sediments in harbours, river mouths, and nearshore areas of the Great Lakes. These in-place pollutants in turn acted as a source of chemicals added to the water and its life. The heavily polluted sediments of Waukegan Harbor in Illinois, for example, contain 500,000 parts per million of PCBs and are one of the major contributors to the PCB contamination of Lake Michigan fish.

Most contamination from groundwater flowing to the Great Lakes comes about through leakage from areas where chemical wastes are buried. At White Lake in Muskegon County, Michigan, for example, researchers found chloroform, trichlorethylene, carbon tetrachloride, and perchlorethylene in the groundwater below property owned by the Occidental (Hooker) Chemical Company.

The magnitude and result of inputs from the atmosphere are only now beginning to be understood. Thousands of chimneys throughout the basin, and in regions beyond, each send out minute quantities of chemicals. Scientists studying Lake Superior discovered that atmospheric deposition accounts for between 80 and 90 per cent of the PCBs entering the lake.

In 1982 scientists found the pesticide toxaphene, used on cotton crops in the southern United States, in lake trout taken from Lake Superior and Lake Michigan. Others also found the same pesticide in fish in a lake on Isle Royale, a large island in Lake Superior. The discovery of toxaphene in this completely isolated wilderness area illustrated the extent to which atmospheric inputs of toxic chemicals had moved into the Great Lakes region. Tests found that as much as 59 per cent of the toxaphene sprayed on fields was lost to the atmosphere – providing the sole source of the toxaphene found in Isle Royale fish.

The 1978 agreement took up control of toxic contaminants as a primary goal, but the diffuse sources of the chemicals and their large numbers made control difficult; and the failure of the governments to enforce existing regulations made the job even harder. In 1983, 43 per cent of Canadian and 18 per cent of U.S. industries were not meeting regulations on effluent discharges. In a quiet understatement, a committee of the Royal Society of Canada / National Research Council investigating the Great Lakes Water Quality Agreement wrote in 1985, "It appears that certain industries continue to be slow in upgrading and implementing treatment processes to bring them into compliance."[14] The International Joint Commission, writing to the two federal governments in December 1984, stated: "Unlike efforts to control phosphorus, there has been limited success in coming to grips with the overall problem of toxics in the

Great Lakes Basin."[15] Nowhere has this failure been more evident than in the area of the Niagara River.

THE NIAGARA RIVER: A CASE STUDY IN TOXIC CHEMICAL MISMANAGEMENT

The Niagara River is home to the highest density of landfill sites containing toxic chemicals in the Great Lakes region. This is the burying grounds of chemicals such as PCBs, dioxin, chlorinated benzenes, and chlorinated phenols. These chemicals, thought to be out of sight and out of mind, slowly leak into the groundwater, working their way into the larger water systems.

The thirty-seven-mile-long Niagara River, connecting lakes Erie and Ontario, is the quintessential bottleneck. The flow of water is crushing, at an average flow of 5,700 cubic metres a second (or 200,000 cubic feet a second), and provides an enormous 83 per cent of the water that travels into Lake Ontario. When, in the early years of the twentieth century, government and business began developing the vast hydroelectric potential of the area, factory after factory competed for space along the U.S. side of the river, seeking to use both the available electric power and the river's vast quantities of water.

As early as 1951 the IJC had noted the discoloration of the river caused by oils, phenols, and various other chemicals. Governments had tackled some of the more obvious direct discharges to the river but much of the problem came from the thorny group of industrial protagonists on the riverside itself. These included a large number of chemical companies, among them Du Pont, Olin, and the Hooker Chemicals and Plastics Company. For over two decades these companies had quietly disposed of the wastes from their production processes in a number of shallow landfill sites scattered around the Niagara area: the 102nd Street Landfill, Love Canal, Hyde Park Landfill, and the S Area, for instance. The wastes buried in these sites included deadly and dangerous chemicals.

The 102nd Street waste disposal site, for example, belonged to both Occidental Chemical Corporation (formerly Hooker) and Olin, which used it as a dumping ground for three decades, until 1971. The corporations buried about eighty-nine thousand tons of chemical waste at the site, including organic phosphates, sodium hypophosphates, inorganic phosphates, BHCs, (including lindane), and chlorobenzenes. Hooker Chemicals and Plastics Company operated the

Hyde Park Landfill from about 1953 until 1975, burying some eighty thousand tons of hazardous chemicals, including calcium fluoride, mercury brine sludge, C-56, organophosphates, acid chlorides, lindane, trichlorophenol, chlorotoluenes, and chlorobenzenes. The landfill also became home to about two tons of the dioxin 2,3,7,8-TCDD.

The buried wastes along the Niagara River had become a relentlessly ticking time bomb, which finally exploded at Love Canal, New York, in 1978 when Hooker Company wastes seeped into the homes of nearby residents. Dr. Beverly Paigen, a medical researcher at the Roswell Park Memorial Institute in Buffalo, investigated the health effects of exposure to the chemicals and found rates of miscarriage three times higher among women living in the area where the chemical exposure was the highest. The rate of asthma was almost four times higher and birth defects occurred three times more often among children born to families living in the high exposure area. The birth defects included extra toes, webbed feet, and kidney abnormalities. The doctor concluded, "The Love Canal is as much a disaster area as any hurricane, earthquake, or flood."[16]

The state government had to evacuate over one thousand residents from the area, many of them permanently. Ann Willis, a Love Canal resident, summarized the horror of the situation in a 1980 poem:

> Take another deep breath. Yes, you feel giddy: your heart races; nausea hits you. It is I, myself, that's been vandalized. *Now* you feel the pain. You want to scream out: you open your mouth and nothing comes out. You open the door of your house and you look up the nice street, and you rush to your car and you cry. Yes, your very existence has been vandalized. You look up and down the street once again; your house is noxious; their houses are noxious; the whole outside is noxious! You want to run! But where? You want to scream! But at whom? I don't want a Love Canal house; I don't want to be a Love Canal victim. But, Oh God, I am![17]

For many, Love Canal became a symbol of the human disaster that can result from corporate carelessness and a misguided belief in an out-of-sight, out-of-mind approach to waste management. But Love Canal was not the end of the toxic nightmare in the Niagara Falls area.

The concentrated industry had led to the establishment of over two hundred hazardous-waste landfills in the area adjacent to the river. But, as well, by 1981 over seven hundred industrial facilities were discharging chemical wastes to sewage-treatment plants, thereby avoiding New York State's pollution permit system. A report on the river pollution by the New York Public Interest Research Group (NYPIRG) stated, "Together, hazardous wastes and industrial discharges constitute an uncontrolled and unprecedented threat to the natural integrity of the Niagara River, the quality of its surrounding environment, and the health and well being of the 380,000 people who drink water from this polluted source."[18] Describing the river as a "sewer," the NYPIRG report blamed industries in the area for polluting the river "almost beyond belief."

The problems did not simply go away by virtue of being recognized. In October 1984 the Niagara River Toxics Committee, composed of representatives from the New York Department of Environmental Conservation, the U.S. Environmental Protection Agency, the Ontario Ministry of the Environment, and Environment Canada, released the results of a three-year study of pollution problems in the river. The report concluded that massive quantities of toxic chemicals were still being discharged to the river each day. The study sampled sixty-nine facilities and found that 89 per cent of the "priority pollutants" came from U.S. sources and 11 per cent from Canadian.[19]

The calculations included only pollutants released from known point sources, and left out non-point sources of chemicals, such as the landfill sites at Hyde Park and 102nd Street. An internal Environment Canada report, leaked to a reporter from the *St. Catharines Standard*, noted:

> The potential for increase in contaminant loadings to the river from non-point sources is very large, since about 215 hazardous waste disposal sites are known to exist in Erie and Niagara counties. Data on *only 15 sites* indicate that they contain about *8 million tons* of contaminated material. If only 1% of this material reached the Niagara River, then 80,000 tons of contaminants will get into the river ecosystem, Lake Ontario, and so on. This compares with the estimated 98 tons per year being discharged now from all *point sources* combined.[20]

The Toxic Committee report concluded, "All the general and

specific objectives of the Great Lakes Water Quality Agreement of 1978 have not yet been attained and are not likely to be attained within the near future."[21]

CORPORATE SPILLS AND LEAKING LANDFILLS

While Love Canal focused attention on the Niagara River, another of the Great Lakes connecting channels – the St. Clair River – was also experiencing its own set of pollution problems. In August 1984 divers hired by the Great Lakes Institute of the University of Windsor were surprised to find a tarry black blob along the river bottom near Sarnia. They got a sample of the substance and sent it off to a laboratory, where tests revealed that it was a highly concentrated mass of perchlorethylene, a toxic chemical used in the dry cleaning industry. Alarmingly, the tarry substance also contained dioxins and dibenzofurans: two cancer-causing chemicals considered harmful even in the smallest quantities.

The next summer, as an investigation of what became known as the blob proceeded, the Dow Chemical Company at Sarnia announced that a mechanical valve failure, between August 13 and August 16, had allowed the release of over forty thousand litres of perchlorethylene to the river. The discovery of the chemical puddles on the river bottom and the subsequent news of the perchlorethylene spill touched off a massive provincial and federal government study of the chemical pollution problems of the river.

The investigation revealed that spills of chemicals into the river by Dow and other companies, including Esso, Suncor, and Polysar, had occurred with disturbing regularity. On February 1, 1985, for example, Dow had released about 4,400 litres of ethylene glycol, a chemical that causes kidney damage. Two months later Dow spilled about 2,500 kilograms of ethylene dichloride, a possible carcinogen, into the river. In May Dow accidentally released to the air about 200 kilograms of a potent carcinogen, vinyl chloride monomer. And later that year, Esso spilled 120 tonnes of slop oil into the river from its plant. Speaking at an October 1986 hearing on Great Lakes pollution in Sarnia, Laurie Montour, a researcher with the Walpole Island Indian Reserve located downriver from the chemical plants, noted that the August 1985 spill of perchlorethylene "was not the most toxic spill nor even the highest quantity, IT WAS THE MOST PUBLICIZED."[22]

These corporate chemical spills were only adding to the toxic chemical burden created by direct discharges from industry and leaking landfills. Sewers attached to the Dow Chemical plant were contributing over 240 kilograms per day of volatile hydrocarbons, including 1,2-dichloroethane, perchlorethylene, 1,2 dichloropropane, carbon tetrachloride, benzene, and ethyl benzene. Effluents from the local Esso chemical and petroleum operations contributed smaller amounts of contaminants, while the discharge from the Polysar plant proved to be a major source of benzene entering the river. Polysar's effluent included something in excess of 110 kilograms a day of benzene. Groundwater leachate entering the river from a Dow Chemical landfill also contained dangerous chemicals, including eighteen parts per million of the pesticide 2,4-D and one part per million of 2,4,5-T.

Toby Vigod, a lawyer with the Canadian Environmental Law Association, expressed the frustration of many citizens. The message that industry in the area had been getting over the years, she said, was that "it is easier to pollute than to clean up."[23] Judith White, a representative to the Lake St. Clair Advisory Committee from St. Clemens, Michigan, expressed a similar frustration at the failure to control the deposits of toxic chemicals. She described the discharges to the river as "the worst sort of filth ... a personal violation and a personal threat."[24]

The problems along the St. Clair River, rather than being unique, proved to be indicative of the continuing difficulty faced in trying to control toxic chemical inputs to the Great Lakes environment. In a remarkably candid but alarming admission, the U.S. Environmental Protection Agency said in its 1986-90 five-year plan, "The sources and role of toxic contaminants are not understood well enough to determine whether the current laws and programs will be adequate for cleanup."[25]

The goal for that cleanup – zero discharge of toxic chemicals – had proved elusive, since controlling chemicals after they had been produced was difficult if not impossible. A more logical approach was to restrict or prevent their production in the first place – at least until companies provided assurances that the chemicals could be used safely and disposed of effectively.

It was no secret that only the most stringent controls would prevent further chemical contamination of people and the environment. In its 1984 biennial report to the governments, the International Joint Commission emphasized the need for a comprehensive strategy to control toxic chemicals. Environmental groups, citizens, and scientists have echoed this need, in no uncertain terms. An environmental activist with Greenpeace articulated the sentiments expressed by numerous citizens around the Great Lakes basin when he told yet another hearing, in Chicago, "Zero discharge of persistent toxic substances isn't idealist pie-in-the-sky. It is a fundamental biological imperative."[26]

LIKE A CANARY IN A COAL MINE

The failure to control toxic chemicals fully is all the more disturbing because there is dangerously limited knowledge of the effects of these chemicals on the environment and human health. What little is known is not good news.

Dr. John Black, a research scientist at the Roswell Park Memorial Laboratory in Buffalo – he is both a cancer researcher and an avid fisherman – has been studying fish to find out more about cancer. During recent years Black has found what appears to be an increasing incidence of cancerous tumors in wild freshwater fish populations – a finding corroborated by other scientists doing similar investigations. He and other scientists believe that in many instances the tumors are related to environmental contaminant levels.

According to Black these cancerous fish are like the unfortunate caged canaries used by early coal miners to warn of the unseen hazards of methane gas. "The fish are warning us of the hazards of the environment," he said, implying that it is time now to stop and take notice before it is too late.[27]

In their study of the fish populations of the Black River in Ohio, Black and his colleagues found that about 30 per cent of the brown bullheads, a bottom-feeding fish, exhibited signs of liver cancer. Another 25 per cent showed signs of skin or lip tumors. Black and his associates also found a high incidence of liver, lip, and skin cancer among bullheads in another highly polluted river, the Buffalo River of New York.

Brown bullheads are only one of the fish exhibiting a high incidence of cancer. Dr. Paul Bauman, another researcher studying cancers in Great Lakes fish, summarized recent findings on possible cancers in salmon:

> Several researchers have reported thyroid hyperplasia (goiters) from coho salmon (Oncarhynchus kisutch) captured in the Great Lakes

> Basin.... A more recent survey indicates that the frequency of overt goiters may be increasing in Lakes Michigan, Ontario, and Erie. During 1975-1976, the incidence of coho thyroid hyperplasia was 6.3%, 47.6%, and 79.5% in these three lakes, respectively.[28]

Dr. Ron Sonstegard, a professor at McMaster University in Hamilton, Ontario, found gonadal tumors in goldfish, carp, and goldfish / carp hybrids pulled from all four of the Canadian Great Lakes. In 1952 Sonstegard caught thirty-eight goldfish and carp hybrids at the mouth of the River Rouge at Detroit; none of the fish contained tumors. In 1977 a survey of the same fish species from the same location revealed a tumor rate of 100 per cent in the older males. Sonstegard, echoing the comments of Dr Black, said the "fish are like sentinels sending out a warning about the presence of dangerous chemicals in the environment."[29]

These warnings were being repeated unknowingly by fish-eating birds. Thomas Erdmans of the Richter Museum of Natural History in Green Bay, Wisconsin, noted that the first deformed herring gull chick had been found on Green Bay in 1973. "At that time," Erdmans reported, "it was a novelty. Today, we have documented five species of fish-eating birds on Green Bay with deformities. This problem is recent and it is getting worse and not better."[30] Larvae of aquatic insects buried in the sediment of the Bay of Quinte, on the north shore of Lake Ontario, were also experiencing increasing incidence of deformities. The rate of deformities had increased over 2,200 times from what the researchers believed was the presettlement deformity rate. The suspected cause was the spread of toxic industrial and agricultural chemicals.

Although more detailed research is doubtlessly needed, the available evidence of increases in fish tumors and aquatic insect and bird deformities is alarming. The evidence is a clear warning, too, that there is even more than the health of insects, fish, and birds at stake.

THE HUMAN HEALTH RISK

In the late 1980s a prominent yellow sign could be seen along the Toronto waterfront. Its bold black letters formed an arresting message: *Check Before You Eat.* The sign, erected by the provincial government, gave fair warning to all people fishing: your catch could be contaminated with toxic chemicals.

Likewise, on the dock at the mouth of the Platte River, a major salmon spawning stream in Lake Michigan, there was also a bright yellow sign bearing silent testimony to the problem of chemical contamination in fish. "Some Great Lakes fish are contaminated with low levels of toxic chemicals," stated the sign. The level of toxic contaminants consumed could be reduced, it said, by using a fileting method to remove the fatty portions of the fish. The sign helpfully illustrated the fileting procedure.

Sadly, signs with similar warnings have become common around the Great Lakes, especially in popular fishing locations. Here and there throughout the region Great Lakes governments have issued advisory notices about the eating of certain fish. All of these notices testify to the results of polluting the waters and the serious implications of that pollution for human health. Is Great Lakes water safe to drink? Are foods grown in the region safe to eat? What are the risks of swimming in the lakes? These are pressing questions for the millions of people who call the region their home. Love Canal aside, the Great Lakes area has proved to be a dangerous place to live. According to a group of prominent scientists from the Royal Society of Canada and the U.S. National Research Council, "Humans living in the Great Lakes Basin are exposed to and accumulate greater amounts of toxic chemicals than humans in other similar large regions of North America."[31] In October 1989 Canada's Institute for Research on Public Policy and the U.S. Conservation Foundation published a 360-page report on the state of the Great Lakes environment; the report noted: "Both wildlife and humans are at risk because of the accumulation of toxic substances in the food web."[32]

To be sure, the levels of toxic chemicals such as DDT and PCBS decreased after the early 1970s, when governments established control programs. But in the 1980s these and a variety of other toxic substances continued to show up in the organisms of the Great Lakes and inevitably made their way into the lives of human beings. To drink water drawn from the lakes, to go into and onto the water for recreation, to just breathe in the air and eat the food from the region: all of these acts became potential and probable means of exposure to toxic chemicals. The proof of such exposure came from new tests identifying PCBS, dioxin, and chlorinated dibenzofurans in the adipose tissue of people living in the Great Lakes basin.

Drinking water taken from the Great Lakes basin invariably contained minute quantities of toxic chemicals. A 1984 City of Toronto report identified eighty-three different chemicals in Toronto's drinking water, including twenty-eight inorganic and fifty-five organic. Some of these chemicals occur naturally in lake water, but many are synthetic. Seven are known human carcinogens; twenty-three are potential human carcinogens. The city's report warned, "There are vast uncertainties about the health effects of many chemicals that have been detected in drinking water."[33] Yet in releasing the report, Dr. Sandy Macpherson, the medical officer of health, told worried citizens that they "can drink tap water with a reasonable assurance it is not likely to cause harm or injury."[34]

The Canadian Public Health Association, after a two-and-a-half year study of Great Lakes drinking water, echoed Dr. Macpherson's confidence. While the association found a number of organic and inorganic contaminants, it noted that the levels exceeded the acceptable standards in only a very few instances.

Despite the low levels of chemicals found in drinking water, studies clearly indicated that there is no safe threshold for chemicals that act as initiators of cancer. The evidence showed that other non-cancer-causing substances can cause toxic effects at very low concentrations. A chemical that performs necessary functions in our body, such as Vitamin B 12, produces positive effects at a concentration of only one part per billion.

In any case, toxic chemicals in tap water are not the major source of chemical exposure for most of us. A preliminary study of non-occupational exposure to toxic substances prepared by Dr. Katherine Davies of the Toronto Department of Public Health found that 85 per cent of exposure comes from food, 11 per cent from drinking water, and 4 per cent from inhalation.

Predictably, fish had become a food source of chemical contaminants. Scientists found that the large predators at the top of the food chain, such as salmon and trout, contained large amounts of chemicals. Indeed, people would have to consume normal quantities of water for more than one thousand years to ingest the amount of PCBS contained in one 500-gram portion of Great Lakes fish.

Other foods also contribute to human exposure to toxic chemicals. Contaminants landing on soil or directly on plants and crops are ingested by livestock or directly by humans. Studies of whole milk from southern Ontario in 1983, for example, found an average concentration of 0.889 parts per billion of PCBS. An intake of one litre of milk a day at this concentration would lead to an annual intake of 327 micrograms a year. Testers regularly detected pesticides and other organochlorine chemicals in beef, pork, eggs, and a variety of other foods.

Katherine Davies again shocked citizens of the Great Lakes basin by announcing at a conference in the summer of 1986 that supermarket-bought food produced in the Great Lakes region, including fruits and vegetables, contained detectable levels of deadly dioxins and furans. Although she did not advise a dramatic change of diet, she recommended precautions such as washing fruits and vegetables thoroughly before eating them.

In the long run there is no doubt about the harmful effects of exposure to low levels of toxic contaminants. While media attention has tended to focus on how chemical exposure can lead to cancer in adults, the greatest threat may well be to unborn or newly born children. The joint Canadian-U.S. study *Great Lakes, Great Legacy?* concluded that "certain identifiable subgroups of the population appear to be at elevated risk of exposure." The subgroups include "human embryos, infants, and children whose parents have accumulated substantial quantities of toxic chemicals."[35] Chemicals are passed to developing embryos through the mother's placenta and to newborn infants through breast milk. Scientists using psychological testing and measurements of newborn infants found, for instance, that smaller birth size, lower gestational age, and neonatal behavioural deficiencies appeared to be associated with infants whose mothers consume Great Lakes fish. These scientists identified PCBS as one possible reason for the results, but pointed out that other contaminants could also be involved.

Clearly, there was a need for more study and effort on the toxic chemical threat in the Great Lakes. Yet there persisted what the Royal Society of Canada and National Research Council called a "serious information gap" – a view reinforced in 1989 by the Conservation Foundation and Institute for Public Policy, who stated bluntly, "The need for more data is urgent."[36] With environmental contaminant residues no longer decreasing, and with new residues such as dioxin being discovered, there remained a "lack of data on contaminant levels in food and hence on total human exposure."[37]

As IJC commissioner Keith Bulen of Indiana wrote in 1985, "The environmental challenge of this decade, and perhaps the remainder of this century, will be understanding and addressing the problem of toxic contaminants in our environment."[38] The battle to control toxic chemical pollution had just begun – with possible dire consequences in waiting for all should the war be lost.

TEN

Fighting Back:

The Movement for a Safe Environment

LEE BOTTS is a fighter who has struggled energetically to halt damage to the Great Lakes environment for over thirty years.

Botts moved to Chicago in the late 1940s and, as a mother of four children, got involved with a neighbourhood group concerned about development of the waterfront region of the city. The U.S. army planned to build a missile base on the Chicago lakeshore, and Botts and others raised strong and practical objections. Ignoring their protests, the army went ahead and built the base, in the process destroying a bird sanctuary and part of the downtown Jackson Park.

Though discouraging, this loss did not deter Lee Botts from getting involved in other Great Lakes issues.[1] A few years later she joined Save the Dunes, an organization formed to preserve the Indiana Dunes at the south end of Lake Michigan. During the late 1960s she became a staff member of the Open Lands Project, a Chicago environmental group. As part of this job she took on the task of uniting Lake Michigan environmental organizations concerned about problems such as power plant construction and pollution. One result of her organizing effort was the formation of the Lake Michigan Federation, which became an important voice of opposition to plans that might endanger the health of the Lake Michigan environment.

The extent of Botts's involvement in Great Lakes issues is

impressive. She has worked for both environmental organizations and government agencies. She helped develop a national energy conservation strategy and, while employed at the Environmental Protection Agency, became involved in developing proposals for a country-wide ban on phosphates in detergents. Her vigour and determination around environmental concerns drew the attention of President Jimmy Carter, who appointed her to the position of Chairperson of the Great Lakes Basin Commission in 1978, a job she held until President Reagan abolished the commission in 1981.

By the mid-1980s, despite setbacks and frustrations with the slow pace of change, Botts remained hopeful about the potential for improvement in the quality of the environment. As she stated, "The main reason I am optimistic about the future of the Great Lakes is the continuing support the public gives to their protection."

But perhaps the main point to be made about her long-time involvement in the region's concerns is that she is not unique. There are countless others like her, people who worry about what has happened – and is still happening – to the Great Lakes environment and who have done a tremendous amount already not only to save it from total self-destruction, but also to bring it back to a proper state of health.

In recent decades the strength and intensity of this opposition have grown by leaps and bounds. The damage done was glaringly obvious. Far-sighted fishers warned of the consequences of overfishing, pollution, and the invasion of non-native fish. Conservationists, naturalists, and farmers decried the destruction of soils, forests, and other vegetation. Scientists and health-care workers pointed out the dangers of continued dumping of sewage and other pollutants into the lakes. In the summer of 1986 environmental activists from Greenpeace plugged the waste discharge system of the Dow Chemical Plant in Saginaw Bay. In a similar expression of concern, angry fishers from northern Ontario took fish killed by a toxic spill at the E.D. Eddy paper plant in Espanola and dumped them at the doors of the Ontario legislature in Toronto.

Although all too often the concerns of these people and groups were ignored, in many important instances their interventions paid off. According to Lynton Caldwell, a U.S. political scientist, "Organized citizen action has nevertheless been the main force behind environmental legislation. Citizen initiative has been important in developing public awareness, in obtaining initial legislation and in amending and administering the laws."[2] The challenge of fighting for a healthy environment has brought new hope that the Great Lakes ecosystem can be salvaged from the cycle of abuse that has existed in the region for the last three hundred years.

THE BATTLE FOR THE NIAGARA RIVER

Like Lee Botts, Sister Margeen Hoffman decided to dedicate herself to the fight for a safe environment and human health. Middle-aged and articulate, Hoffman became one of the most vocal and effective crusaders working to stop the deadly flow of chemical contamination into the Niagara River. In 1979 she became the Executive Director of the Ecumenical Taskforce of Niagara Falls, N.Y., an interfaith group organized to address community problems arising from chemical and radioactive wastes.

A number of local churches formed the group in response to the tragedy of Love Canal, offering assistance to the families displaced by the disaster, providing food, shelter, and spiritual and emotional support to the victims. The very establishment of the task force proved to be a significant step. According to Hoffman, "Although the churches are often very quick to respond to natural disasters, Love Canal was the first time they had ever responded to a technological disaster."[3]

For Sister Margeen Hoffman, "Love Canal was a harbinger of many things to come." Besides revealing the horrors of the "out-of-sight-out-of-mind" approach to waste management, the Love Canal episode also led to a new focus on other local pollution problems. Shortly after finding out about the health hazards of the chemicals buried at Love Canal, the Ecumenical Taskforce and the residents of the Niagara frontier received further confirmation of the short-sighted nature of past waste disposal practices. They began to see "that Love Canal was only one phase of the thing."

As Hoffman said, "We found out right on the heels of Love Canal – not even on the heels, but alongside it, parallel to it – the Hyde Park dump contained even larger amounts of waste material." In a situation similar to Love Canal, the buried chemicals at Hyde Park were seeping slowly out of the landfill into the Niagara River. The deadly chemicals entering the river included a portion of the over two thousand pounds of dioxin buried at the site – dioxin that, according to Hoffman, was "flowing, flowing every day."

The Ecumenical Taskforce has employed a number of tactical

weapons in its struggle to stop the contamination of the Niagara River and surrounding area, including lawsuits, government lobbying, research, and public presentations. Much of their inspiration comes from Biblical prophets, people who saw what was happening and dared to talk about it. Like the prophets, the Ecumenical Taskforce raised a subject that most people would just as soon forget – in this case, chemical contamination of the environment. "I'm not going to win any awards for Miss Tourism Niagara," Hoffman said, "but people are now talking about these issues – people in business who not too long ago certainly wouldn't. Every time we do this we are able to get a little more leverage. There are now some new laws and regulations."

On the Canadian side of the river, Sister Margeen Hoffman's counterpart is the flamboyant Margherita Howe of Niagara-on-the-Lake. Howe heads the group Operation Clean Niagara and first started fighting the pollution of the Niagara River in 1979 when she heard about the plans of the SCA Chemical Company to discharge one million gallons of treated chemical waste into the river each day. The drinking water intake for Niagara-on-the-Lake, she discovered, was only two miles downstream from the proposed discharge. This news led to two hundred area residents attending a company information meeting on the plan to discharge the effluent. Two weeks later these local people formed Operation Clean.

The group appeared at hearings on the proposal and made its opposition known. Although the company eventually got the go-ahead for the treatment plant, as a result of the citizens' actions the state imposed restrictions on the quality of the effluent and established a vigorous monitoring program.

After becoming active on the issue of water quality in the Niagara River, Operation Clean asserted itself as a watchdog on pollution. Together, Operation Clean, the Ecumenical Taskforce, and the Toronto environmental group Pollution Probe took the bold move of intervening in a settlement of a lawsuit between the U.S. federal government and the Hooker Chemical Company over contamination from the Hyde Park dump. The groups did not believe that an agreement for containment of the dump that had been reached between Hooker, the EPA, and the State of New York would solve the problem of the leaking wastes, and issued a legal challenge to the agreement.

After presenting their preliminary position to the judge, the groups were granted "friends of the court" status, which allowed them to argue the adequacy of the agreement. They presented evidence showing that the wastes, which included dioxin, would continue to leak and that the only prudent action was the excavation and incineration of the waste. Although the judge eventually ruled that the original containment proposal was satisfactory, the trial focused media coverage on the issue, and pressure increased on both governments and industry to deal more effectively with the problem of the leaking dumps in the Niagara area.

Soon after, another of the notorious Hooker dumps, the S Area, was the subject of yet another court proceeding over cleanup. Once again the Ecumenical Taskforce, Operation Clean, and Pollution Probe lobbied for a thorough and effective program to prevent the wastes from reaching the river and the Niagara area environment. As a result the technical requirements for cleanup of the S Area became much more stringent than those for the Hyde Park settlement. Rick Findlay, a senior Environment Canada official responsible for the Niagara River, underscored the groups' effect: "The actions of citizens have been a tremendously important factor in prodding governments to examine and respond to the Niagara River problems."[4]

There are countless more examples of individuals and groups throughout the Great Lakes Basin who have reacted to problems of pollution in their local areas. Members of the Akwesasne Indian Reserve near Cornwall, Ont. and Messena, N.Y. began a tireless and extremely effective campaign to stop the discharge of PCBs and flouride from three of the world's largest companies, General Motors, Alcoa, and Reynolds Metals. Along another of the basin's major connecting channels, the Detroit River, the Windsor and District Clean Water Alliance took on the Detroit Sewage Treatment Plant, using both the courts and media exposure to fight against its contribution to pollution. Citizens For a Better Environment in Chicago actively campaigned for tighter pollution control standards at Lake Michigan steel mills and intervened in a State of Michigan permit hearing over the pollution discharge from the Dow Chemical plant in Midland, Michigan. The Michigan United Conservation Clubs and the Save Lake Superior Association took up the challenge of stopping the spread of chemical contaminants.

Many organizations took their concerns to federal, provincial, state, and local governments to get passage of laws that would further protect the Great Lakes environment. The Sierra Club, one of the largest environmental organizations in the region, effectively injected the issue of air toxics into the congressional clean air debates of early 1990. The Canadian Institute for Environmental Law and Policy and other Canadian groups developed an aggressive campaign to strengthen Ontario's water pollution laws. At the local level a committee of concerned citizens pushed Buffalo city council to pass a resolution that will require the city to adopt an Integrated Pest Management Program and reduce the city's use of pesticides.

Workers, many of them confronted by chemicals in their workplaces, became active in the fight against chemical pollution. Bill Van Gaal of the Canadian Auto Workers environment committee told a hearing on Great Lakes pollution in October 1986, "We realize that pollution of the workplace is directly related to environmental pollution.... We are intent on eradicating both for the well-being of mankind."[5]

WETLANDS REVISITED

There is no denying that massive human activities – of settlement and industry, of a certain kind of economic and social survival – must inevitably alter the landscape, and sometimes do so necessarily. It is impossible to imagine a modern southern Ontario or New York state where elks could still wander freely or a pair of bald eagles could still perch on trees every few miles along the shoreline. But by the later twentieth century, with forests cleared, swamps drained, dunes bulldozed, farming land turned into urban housing, and waters heavily polluted, there was also no denying that we had completely broken down the original components of the ecosystem.

It was equally clear that this development could in no way be construed as being in the present population's best interests – or morally acceptable. At the same time, each successive environmental loss further undermined the ecosystem's perilous state of health.

The loss of wetlands is a prime example. Wetland areas are essential to the health and well-being of the environment: they act as a flood storage area, as a natural filter to protect the water of the Great Lakes from nutrients and toxic materials, and as habitat and nursery areas for fish, mammals, and birds. The Lake Erie and Lake St. Clair marshes, for example, provide the most important waterfowl staging area south of James Bay. About two-thirds of the North American population of tundra swans still use these marshes during migration.

Despite their important role in the ecosystem, wetland areas have fallen by the wayside in the march of progress. In Michigan over 70 per cent of the original wetland areas have been converted to other uses. In Ontario the loss in the Great lakes portion of the province has been staggering: 77 per cent of the wetlands present when the settlers arrived have disappeared. This loss has several serious repercussions: the reduction of wildlife populations, the disruption of local water tables, and increased flooding. On Lake St. Clair, for example, the numbers of ducks observed in the spring of each year declined by a huge 79 per cent between 1968 and 1982.

One of the last remaining wetland areas on the north shore of Lake Ontario is Oshawa Second Marsh, about thirty miles east of Toronto. Second Marsh, a 290-acre cattail marsh with an associated upland forest, is a major migration stopover point for waterfowl, and has long been one of the top five locations for banding of migrating ducks in the Atlantic flyway.

Despite its uniqueness, the marsh was threatened by urban encroachment beginning in the 1970s. The Oshawa Harbour Commission, which owns the land, put a stop to the banding operations in 1973 by restricting access of banders to the marsh. The commission, it seemed, considered the site more ideal for a deepwater harbour than for a bird resting stop, and made plans to dredge the marsh, blast the bedrock, and open the surrounding area up to industrial development.

For over ten years a battle raged to determine whether the marsh would be preserved or destroyed. Fortunately for the marsh and the Great Lakes ecosystem, a dedicated group of people worked hard to ensure its continued existence. Jim Richards, a night-shift worker at the General Motors car plant in Oshawa, was involved in that movement. With a basic passion for birds, Richards was drawn to the Second Marsh and got to know the rich birdlife there. He also came to know that the great biological productivity of wetland areas such as Second Marsh is essential to maintaining the already dwindling wildlife populations.

The area became a special place for Richards. He and others with the same interests fought to preserve the marsh and prevent its development into a harbour of questionable value. Organized under the umbrella of the Second Marsh Defence Fund, they organized demonstrations and letter-writing campaigns. Their aim, according to Richards, was to see the marsh used in "a very passive way; as a place to instill in children a deep sense of appreciation for wetlands."[6] By the late 1980s the group's tactics and long struggle had continued to prevent the destruction of the wetland.

Farther west, at the south end of Lake Michigan, sand dunes were similarly under attack, including some of the finest formations in North America. Organized in 1952 as the Save The Dunes Council, a dedicated group of women took up yet another long struggle for the preservation of a unique ecological feature. In the late 1960s the group successfully lobbied Congress to purchase some of the dunes and create the Indiana Dunes National Lakeshore. More recently the group focused its efforts on encouraging the government to incorporate additional dune lands into the park and on ensuring effective management of the park lands. The group became engaged in an almost non-stop fight with the federal government over the allocation of funds for enlargement and maintenance of the park. But members have a wider interest as well. As Charlotte Read, executive director of the council, noted, "We are increasing our attention to threats to the National Lakeshore from sources outside the park, such as air and water pollution."[7]

Such groups and individuals play a crucial role long abdicated by government and industry – a role for which future generations will doubtless be grateful. They keep alive the hope that the valuable natural resources of the Great Lakes region can continue to contribute to the ecological health of the region. But battling to preserve and protect specific natural areas within the Great Lakes basin can only be effective if damaging practices throughout the basin are stopped. Because changes in one part of the basin invariably influence conditions in other parts, scientists and activists working to protect the lakes and the surrounding environment have come to appreciate the importance of the larger interrelationships. Efforts to protect the Great Lakes, they know, must be co-ordinated and linked.

Related to this is the historical and geographical fact that the lakes are an international resource shared by the citizens of two countries. The expanding cross-section of people working to preserve and protect the Great Lakes environment has increasingly recognized the importance of co-operative action across the international border.

GREAT LAKES UNITED AND INTERNATIONAL ORGANIZING

"United! Together we can protect and promote our common treasure – the Great Lakes Basin." This is the rallying cry for Great Lakes United (GLU), an international coalition of environmental groups dedicated to preserving and protecting the Great Lakes and the St. Lawrence River.[8] Formed in 1982, GLU soon achieved remarkable success in building an effective international coalition and its broad membership base brought added strength to the voices expressing concern for the welfare of the Great Lakes region. Michigan Governor James Blanchard called the GLU "informed, effective and influential."

A former GLU executive director, David Miller, said the organization was striving to bring about "co-ordinated citizen action on important public policy issues affecting the Great Lakes."[9] Michigan United Conservation Clubs, United Auto Workers, Pennsylvania Sportsmen Federation, Canadian Environmental Law Association, and La Société pour Vaincre la Pollution all joined up to be among the over 180 Canadian and U.S. members.

Funded by foundation grants and membership fees, GLU has tackled such issues as winter navigation, Great Lakes water diversions, and chemical contamination of the lakes. In 1985 it identified toxic contamination as "the single greatest threat to the Great Lakes."[10] The group participated in lobbying to ensure re-authorization and strengthening of the U.S. Clean Water Act, and continued funding of the Superfund for hazardous waste cleanup. It also encouraged federal, provincial, and state governments to "promote alternatives to the generation, land burial and discharge of toxic hazardous substances."[11]

One of GLU's important strengths, according to Miller, is that it brings "a broader ecosystem perspective to local concerns." GLU has, for example, submitted briefs and appeared at hearings in support of the concerns of groups of citizens along the Niagara River who are worried about the toxic chemicals leaking from landfill sites into the river. GLU also announced early in 1986 that it was joining with New York State environmental groups in filing a lawsuit against a Niagara River industrial association that had petitioned the government for a relaxation of existing industrial discharge permits. In announcing its decision to participate in the legal action, GLU stated, "Strong New York State water quality standards and laws are an essential element of protecting the Great Lakes ecosystem."[12] GLU and its allies won the court case and GLU reported to its members: "Great Lakes United hopes that this sound victory in the New York case will provide legal precedents to help other jurisdictions in their quest for improved Great Lakes Water Quality."[13]

GLU has also used its strength as an international coalition to pressure the U.S. and Canadian federal governments. In 1985, 1987, 1988, and again in 1989, over twenty people from GLU member groups spent a week in Washington lobbying elected representatives and U.S. government officials for a stronger U.S. federal commitment to the protection of the Great Lakes basin. The group joined with other Canadian environmental organizations to develop and submit a "Greenprint for Canada" to Prime Minister Mulroney in the summer of 1989.

Among GLU's most successful activities was a series of public hearings held around the basin during the summer and fall of 1986. GLU staff and board members visited nineteen Great Lakes cities, from Duluth in the west to Montreal in the east. The hearings, designed to provide citizens with an opportunity to express their views on Great Lakes issues, were an overwhelming success. Over twelve hundred people attended and the meetings became important forums for raising local concerns about Great Lakes issues. They helped to develop further public interest in protection of the ecosystem and in taking action. Throughout the tour the hearing panel heard about and observed many of the environmental problems facing the region: plumes of black smoke, old dumpsites leaking chemicals, and agricultural runoff releasing pesticides and phosphorus to the lakes. But still, John Jackson and Tim Eder, two of the tour organizers, came out of all this with positive feelings:

> We returned from our tour of the Lakes full of hope. This hope is based on our faith in citizens' actions. We were inspired by the determination and commitment of citizens throughout the Basin to find and implement solutions to the problems.[14]

These actions, Jackson and Eder concluded, were "having dramatic impacts."

The formation of the GLU – with its premise that environmental issues do not respect political boundaries – illustrated the enormous potential of international co-operative action for combating environmental problems. This point was further emphasized by the establishment of the Centre for the Great Lakes, another international organization dedicated to developing programs to manage and conserve the region's natural resources more effectively. From its inception in 1983 the Centre for the Great Lakes, with offices in Chicago and Toronto, has served as a catalyst, bringing together individuals and groups committed to seeking practical solutions for Great Lakes problems. Its superb research materials have contributed new information and understanding on important issues. The conviction behind the group's formation, according to its publicity brochure, was that the "environmental and economic well-being of the lakes and the surrounding basin are inseparably linked."

By the late-1980s the activities of groups such as the Save Lake Superior Association, the Ecumenical Taskforce, Operation Clean, the Second Marsh Defence Fund, Great Lakes United, and the Centre for the Great Lakes, as well as the work of hundreds of other groups and individuals to preserve and protect the Great Lakes, suggested strongly that further deterioration could be halted. As Glenda Daniel of the Lake Michigan Federation remarked, "We've come a long way in the last 20 years, but there is much more to be done – work that won't be done unless citizens remain active and interested."[15]

In every corner of the basin, people had responded to the continual abuses of the lakes with concern that often turned into anger. They had begun to organize, to research, and to act to stop such mistreatment. Many had become convinced of the need for a new ethic and philosophy to guide our relationship with the Great Lakes environment. Clearly, the ecosystem could no longer be taken for granted.

PART THREE

ELEVEN

Restoring Health to an Ailing Ecosystem

IN SPRING 1987, while birdwatching near Point Pelée on Lake Erie, I watched a young eagle fly gracefully over a marsh in the cool light of a morning sunrise. I could just make out a coloured wingtag identifying the bird as one that had been re-established in the area through a government program. Though it lacked the spectacular white head and tail of an adult eagle, its size was awesome, its flight powerful: all in all it made for a surprising and captivating sight.

By the late 1980s, the bald eagle – a symbol of the problems of the Great Lakes in the 1960s – had become a symbol of the potential to restore the ecosystem to health. Its populations had been decimated during the 1950s and 1960s when its traditional nesting habitat had been destroyed and the birds and their eggs had been poisoned by deadly toxic chemicals, especially DDT. But the dedicated efforts of wildlife officials and naturalists in both Canada and the United States combined with the partial control of toxic chemicals gradually brought eagles back to the Great Lakes region. By building artificial nesting platforms, wildlife officials in Michigan successfully induced bald eagles to nest in parts of their former range. The first sign of success came in 1977 when five young eagles flew from two nests constructed on artificial platforms. Eagles also began returning on their own to nesting sites on Lake Superior.

More recently, officials used a modified falconry technique known as hacking to re-establish eagles in some of their former haunts in New York State and southern Ontario. Wildlife workers took flightless young eagles from their nests in areas where they were abundant, such as Alaska and northern Ontario, and transferred them to artificial nests in the Great Lakes region, to be fed by human parents. The hope was that when these young eagles matured they would return to nest in the areas where they were raised.

Eagles can now be seen in districts where they had been absent for years, and natural nesting has increased. The preliminary success of this rehabilitation program – though its long-term survival is not certain – serves as an important symbol of the potential to undo some of the damage that has been done.

But, like a diseased patient, the Great Lakes environment needs more than an end to the spread of the illness. It needs a strict and systemic program of recovery and restoration. In the early 1970s a group of university-based scientists made such a proposal, for a program of ecological rehabilitation. Their goal was simple: to stop and reverse the overall long-term trend towards ecological degradation and restore some of the healthier features of the ecosystem, such as lake trout, wetlands, and mature forests, elements that had been lost or reduced. One of the scientists, Dr. Henry Regier, stressed that people in the region needed to get away from simply "reacting to the bad" – they needed to involve themselves in "recovery and resurgence of desirable things."[1]

The idea of ecological rehabilitation, though far from new, was slow to catch on with government agencies. In 1976 yet another meeting, of Canadian and U.S. academics and scientists, proposed the formation of an international reference group on Great Lakes Rehabilitation and Restoration. The group concluded that pollution abatement programs alone were not going to bring the Great Lakes back to health: "The natural recovery processes which will occur as water quality objectives are reached will need to be augmented by active programs of rehabilitation and restoration."[2] The meeting emphasized that "rapid progress could reasonably be expected once both countries agree to encourage and support it."

These academics and scientists presented their proposal to both the International Joint Commission and the Great Lakes Fishery Commission (GLFC). Despite the obvious relevance of the idea to the

IJC, it was the Great Lakes Fishery Commission that initially endorsed the idea and undertook further research on how to implement rehabilitation strategies. The commission funded a study of the concept, which led to publication of a report, "Rehabilitating Great Lakes Ecosystems," in December 1979.[3] The report concluded that it was technically feasible to design and implement ecosystem rehabilitation strategies in the region.

THE AREAS OF CONCERN

There is no formula for ecological rehabilitation. The potential strategies depend on the stresses placed on the ecosystem and on the factors preventing its natural rehabilitation. For instance, the reproductive success of the walleye has been reduced by the destruction of its habitat and the covering of its spawning beds with silt. People, especially sports fishers, interested in rehabilitating the walleye along the north shore of Lake Erie and in other Great Lakes locations have therefore focused their efforts on restoring the original conditions in former spawning areas and on creating new habitats. Similarly, the re-establishment of a healthy forest is often hindered by the lack of available natural seed in cutover and burned lands. Planting, soil management, and other forest management techniques are therefore needed to rehabilitate areas where forest regeneration has not occurred, or where it has occurred so that only some of the forest species, such as aspen, can survive.

Each degraded area in the Great Lakes requires specific rehabilitation strategies designed especially for the stresses placed on that area: this idea appears to be catching hold. In 1972 the Great Lakes Water Quality Board warned the International Joint Commission that at least forty-two areas in the region were badly hurt and needed attention. The IJC defined these as "areas of concern," and thereafter the commission and the public alike spent long years witnessing a frustrating and continued failure to deal fully with the problems. In its 1984 biennial report to the governments, the IJC concluded somewhat dryly, "Because of the apparent lack of progress in resolving the problems identified in the areas of concern ... major effort must be made to correct these situations."[4]

In response, the Water Quality Board requested in its 1985 report that various agencies responsible for the forty-two identified areas of concern draft what they called remedial action plans. The board asked that the plans define the environmental problem, determine the geographic extent of the area affected, identify all the beneficial but impaired uses of the area, describe the causes of the problems, and outline remedial measures to resolve problems and restore beneficial uses. In addition the board asked the agencies to develop a schedule for implementation and completion of remedial measures, and to identify the jurisdictions and agencies responsible for implementing them. Two of the locations that took up this planning were Green Bay, Wisconsin, and Hamilton Harbour, Ontario.

GREEN BAY: A CASE STUDY IN REHABILITATION

Situated on the western shore of Lake Michigan, Green Bay contains some of the most productive and intensively used land in the Great Lakes basin.[5] About one-third of the total drainage area of Lake Michigan flows into Green Bay.

Despite its position at the western end of the basin, Green Bay was one of the earliest areas explored. The French missionary Father Marquette and the explorer Jolliet travelled by canoe to the bay in 1673. Marquette, in later years, established one of the first Great Lakes missions there.

Fourteen rivers and numerous tributaries drain into Green Bay, the largest being the Wolf-Fox River system. Although about one-third of the total watershed is forested, much of the rest is intensively farmed or occupied by urban development. There is heavy industrialization along the Fox River valley, which contains the largest concentration of pulp and paper mills in the world.

The environmental problems of lower Green Bay and the Fox River did not occur overnight: they took over a century to develop. One of the greatest losses was extensive coastal marshes. During the 1840s, fifteen square miles of coastal marshes and seventy-two square miles of coastal swamps existed along Green Bay's western shore. Within the past century 60 per cent of these marshes were converted to agricultural land, filled with dredge spoils, or invaded by cottage settlements. Only six square miles of marsh and twelve square miles of swamp remained during periods of high water levels. Damage to the once diversified fishery of Green Bay became so extensive that by the 1980s only the yellow perch was harvested in significant quantity.

Among the early human activities most destructive to the bay was the discharge of wastes from logging and, later, pulp mill operations. By the early twentieth century, the huge white pines were gone and the pulp and paper industry had taken over as the major user of the Green Bay forests. Its mills flourished along the lower Fox River. Industrial waste discharges kept pace with the expansion.

The environmental awakening of the early 1970s led to some controls and improvements, but the Green Bay ecosystem remained in ill health. Responding to the conditions, researchers from the University of Wisconsin Sea Grant Institute and the Wisconsin Department of Natural Resources began a concentrated effort in 1978 towards restoring the Green Bay ecosystem. The scientists got funding from the University of Wisconsin Sea Grant and the Great Lakes Fishery Commission to develop a rehabilitation plan, and they held a series of workshops, bringing together the users of the bay and government officials responsible for its management. They felt that by getting both the public and the resource managers involved, they could all develop effective and workable policies for rehabilitation of the bay.

A preliminary plan identified the critical stresses acting on Green Bay and suggested a number of actions: establishing a 75 per cent reduction in the level of phosphorus in the bay; implementing non-point nutrient control measures for the watershed; adding monitoring programs for toxic chemicals; constructing barrier islands in the lower bay to reduce waves that resuspended nutrients into the water column; and enhancing the bay fishery. In particular, the plan proposed controls on the non-native carp, a species both damaging to aquatic vegetation and disruptive to other fish.

The team hoped that the plan and the workshops would set in motion a process leading to the further refinement and eventual implementation of strategies for rehabilitation – which in fact did happen. The report led to the formation of a regional task force including a broad spectrum of representatives from government agencies, public institutions, and the private sector, co-ordinated by an existing regional planning agency. Calling itself "Future of the Bay – Technical Advisory Committee," the group moved to establish the first annual "Bay Awakening," held in October 1982. The conference served as a stepping stone to the rehabilitation of Green Bay. Various activities – boat rides, tours, exhibitions, demonstrations – greatly expanded public awareness of the bay's problems and of possible strategies to improve its condition.

The tours, workshops, and displays all contributed to increasing public concern for building a new relationship between humans and the Green Bay ecosystem. The approach especially illustrated that rehabilitation of the Great Lakes required more than a refinement of pollution control technologies and the reintroduction of desirable species. It required new ways of thinking about the environment and the human relationship to it.

The success of the initial conference led to another, in November 1983, this one focusing on the difficult question of what to do with the chemically contaminated sediments in the bay. The conference gave its support to an Army Corps of Engineers proposal to deal with contaminated dredge material.

The expanding interest and enthusiasm for the cleanup of Green Bay led to public interest in the development of a detailed remedial action plan for submission to the IJC. The key to the new plan, according to Bud Harris, one of the original motivators of the ecosystem rehabilitation idea, was that the agencies that had to do the work had now become active in the process. Harris pointed with pride to the public advisory group composed of over thirty people who ensured that the plan reflected the interests of the public. Given that restoring the Green Bay ecosystem to health would take time, Harris for one was confident that at least there was a process in place to undo some of the damage done. Another environmental activist, Rebecca Leighton, summarized the viewpoint of the participants: "It's a monumental effort to get so many people working together and communicating, but the resulting plan will be an important tool for the clean up of the Fox River and Green Bay. It's worth it."[6]

The plan appeared to have enormous potential for stopping environmental abuse. It had put in motion a process that would force officials to address problems long avoided. In September 1987, looking out onto a greatly improved Fox River, Harris said, "People are beginning to look at the river as something other than a sewer. Now they like living along it."[7]

HAMILTON HARBOUR: PLANNING FOR REMEDIAL ACTION

Forming an almost perfect triangle at the western end of Lake Ontario, Hamilton Harbour is separated from the lake by a narrow sandbar pierced in the middle by the Burlington Ship Canal. The canal allows ship traffic to move between the harbour port facilities and the open waters of Lake Ontario.

Hamilton is a steel town. The lake freighters entering the harbour are usually filled with iron ore and other raw materials heading for the mills of Dofasco and Stelco on the south shore. Along the north shore, housing and parkland are the predominant land uses. Agriculture is also important in the 500-square-kilometre drainage area of the harbour. The combined effects of industry, agriculture, and urban development have led to the harbour's designation as an area of concern.

Like in Lake Erie in the mid-1960s, the most significant ecological problem in the harbour is the lack of oxygen, a problem created by the breakdown of ammonia discharged from industry and sewage-treatment plants and by the large phytoplankton biomass caused by phosphorus pollution. But oxygen depletion is only one of the problems plaguing the harbour. Toxic chemical contamination and bacterial pollution, as well as extensive destruction of wildlife habitat, have all contributed to the area's malignancy. Although pollution control programs implemented through the 1970s had brought some improvements to environmental conditions, the harbour remained a highly polluted body of water underlaid with excessively contaminated sediments.

The Ontario Ministry of the Environment initiated a remedial action plan for the harbour in July 1986. To develop this plan, the ministry established a writing team with representatives of the major agencies responsible for managing the harbour. In addition it created a stakeholder group comprised of representatives of agencies, organizations, government departments, industry, and citizens' groups. All of these bodies had either made use of or had jurisdiction over the use of the harbour. In total forty-nine people joined the stakeholder group, including an alderman of the City of Hamilton, a member of the Bay Area Residents' Association, officials with Stelco and Dofasco, and yacht club and naturalist club representatives.

Fortunately, the participants became engaged in the process with a sense of hope and enthusiasm for improving the harbour. The Hamilton Naturalist Club, for example, stated that it "firmly" believed that the Remedial Action Plan for Hamilton Harbour was a "tremendous opportunity to protect, conserve and improve the Harbour environment."[8] At regular meetings the stakeholder group developed a collective vision of what it wished to see Hamilton Harbour become through the process of rehabilitation. The stakeholders agreed: "That the water quality in Hamilton Harbour's drainage basin be improved to ensure continued human health and well-being."

One of the most significant and demanding of the goals identified by the stakeholder group was that of re-establishing a "naturally reproducing warm water fishery in the Harbour." Pike, bass, and a variety of other fish had once been common in the shallow marshes of Hamilton Harbour. The message of the people of Hamilton was clear: they wanted edible fish back in their water.

The first steps to reintroducing the fishery included surveys of fish spawning habitat and radio tagging of adult pike released into the harbour. In addition, researchers from the Canada Centre for Inland Waters undertook an experiment to speed up the process of recovery for the harbour. They began pumping oxygen into the waters to help break down some of the wastes accumulated over the years on the bottom of the harbour.

While the stakeholders were committed to this oxygen pumping scheme, they also made it clear that they did not want to see oxygenation as a substitute for stopping the waste inputs – demonstrating their commitment to stopping the problems of pollution at their source. When a proposal to divert sewage wastes from Hamilton Harbour to Lake Ontario was raised, the members of the group were adamant in their abhorrence of this as a "cleanup option." Summarizing the sentiments of the stakeholders, one member of the group explained, "Dilution never has been and never will be a solution to pollution."

The "phase one" report of the stakeholders and the writing team, presented to the IJC in the fall of 1989, was not to be the end of the rehabilitation process. A committee was also formed to ensure implementation. As Gil Simmons, a harbour resident and stakeholder representative, noted, "Rehabilitation of the Harbour is an ongoing process that will require continual work by everyone who has a stake and interest in the harbour."[9]

THE REHABILITATION OF BLACK CREEK, AND MORE

The restoration of large degraded components of the Great Lakes ecosystem, such as Green Bay and Hamilton Harbour, can in turn have a beneficial influence on the overall health of the Great Lakes environment. But smaller scale projects are also possible and can have equally important effects on local areas. One such project attempted to restore the health of Black Creek, one of the many tributaries that flow into the Toronto waterfront.

Black Creek is a medium-sized tributary that winds its way

through the heart of Toronto from its source in a glacial moraine just north of the city. Like all the tributary streams in the Toronto area, it has endured a succession of abuses. The stream has been subjected to chemical and bacterial pollution and has serious erosion problems along its banks. Its waterflow suffers from ecologically unhealthy and dangerous extremes. Its water quality problems have also contributed directly to the contamination that has forced the closure of some of the Toronto beaches to swimming.

Stating "Black Creek is at a pivotal point," a coalition of citizens, politicians, and university researchers began work on a plan to rehabilitate the creek.[10] They hoped this rehabilitation would serve as a model for restoring other Toronto rivers and streams and offer proof of the viability of ecological management. The alternative was unacceptable: to watch Black Creek "degrade further into a slimy trickle in summer with violent spates in spring and fall."

Much of the stress on the creek grew out of the extensive urbanization of the surrounding area. Large pieces of nearby land were paved over, causing water levels in the creek to rise rapidly in wet weather. Many of the former creek valleys, which had previously served as floodwater reservoirs, were filled in by housing developments, removing much of the creek's natural defence against flooding. Local governments had also constructed concrete channels down sections of the creek to help move the water more quickly to Lake Ontario. This channelling destroyed much of the natural habitat along the stream and damaged the value of the creek as a wildlife refuge. Direct sunshine on the water in these concrete tubs raised the water temperature to levels above those tolerable to most warm water fish. Finally, the erosion of soil from nearby construction sites and the washing of street dusts into the creek added large amounts of sediments.

The coalition of people concerned about the creek hoped to reduce these abuses to the point where natural ecological recovery could occur. In addition, they argued that some natural features of the stream would need to be reconstructed. They planted shade trees along parts of the creek and worked to secure vegetation cover for parts of the creek valley that had been eroding. In the long run, they said, strategies such as replacing gabions and concrete channels would also have to be undertaken. It was possible that fishing could be reintroduced, "when contamination is at acceptably low levels."[11]

To undertake further work on the rehabilitation of the creek, the ad hoc group attempted to secure money from the City of Toronto and called for the creation of a task force comprised of citizens and appropriate agencies to design and carry out a rehabilitation plan. This kind of enthusiasm – with the proper support – only added to the realization that Black Creek and other bodies of water like it could indeed be restored to a condition of health.

A hundred and fifty miles north of Black Creek, on a tributary of the Sydenham River that eventually flows into Georgian Bay, the members of the Sydenham Sportsman's Association are the instigators and practitioners of a project to restore fish to a natural channel of the river. Club member Osbourne McArthur suggested the project. One day in 1984 McArthur saw forty pairs of rainbow trout in the river. A week later he found the stream totally dried up, along with hundreds of dehydrating trout eggs. He and other members of the club initiated a rehabilitation project involving sixty men working weekends and evenings to create a steady flow of clear cool water through the channel. With a small amount of funding from the Ontario Ministry of Natural Resources, they improved the stream bed by placing small pebbles and rocks on the bottom to improve the habitat for the trout. "Every little rock provides a niche that will provide a food base for that little trout to mature," club member Fred Geberalt said.[12] The club also built a small dam to guarantee a steady flow to the channel from the numerous spring-fed rivulets in the area. They made plans for a fish hatchery to stock the stream.

Other volunteer organizations raised funds to undertake similar rehabilitation projects in other parts of the Great Lakes basin. On one drizzling Saturday morning in June 1986, in a cleanup effort organized by the Friends of the Rouge River and assisted by the works crews of fourteen sponsoring communities, over two thousand people turned out to pull garbage – including shopping carts and auto parts – out of the river. A similar effort along the Buffalo River in May 1989 removed piles of garbage, including a number of rusty cars, from the river and its banks. Again, the efforts of these groups prove – if any proof is necessary – that the restoration of the Great Lakes ecosystem is desirable, possible, and worth working for.

REHABILITATION: FROM LAKE TROUT TO TURKEYS

Lake trout, originally the major large predator fish in the Great Lakes, is ideally adapted to the lakes, and one of the most intensive rehabilitation efforts underway in the Great Lakes region is the restoration of lake trout to a point where stocks become self-sustaining. The ravages of the lamprey decimated trout populations during the 1940s and 1950s, but chemical controls on the lamprey starting in the 1960s opened up the possibility of restoring the lake trout to its former position of prominence. Fishery managers throughout the region have since pursued this goal.

By the late 1980s over 125 million fingerling / yearling lake trout had been released in the lakes in hopes of re-establishing the species. Stocking has taken place in Lake Superior since 1958, in Lake Michigan since 1965, in Lake Huron and Lake Ontario since 1973, and in Lake Erie since 1975. The rehabilitation seemed simple. The idea was that the hatchery-reared fish would survive in the lakes until they sexually matured at the age of six to eight years, and would return to their historically successful spawning sites. But what appeared simple at the outset turned out to be much more complex. The fish grew well in the lakes but did not return to the spawning sites. Instead they returned to the areas where they had been released, and had little success spawning at these sites.

Fishery officials responded by changing the stocking program, planting trout in areas where the opportunity for reproduction seemed greater. And some natural reproduction of the trout did occur. Now, in remote areas of Lake Superior, up to 90 per cent of the existing lake trout are naturally produced. Wild lake trout fry are now being found in Lake Michigan and fry and wild yearlings have been collected from Lake Huron.

Despite these encouraging signs, the rehabilitation of the lake trout has proceeded slowly. Factors other than the choice of spawning sites hindered the fish, causing scientists to speculate about the competition, especially young alewives and smelt competing with young trout for food. They also found that adult smelt might be acting as predators on the young trout. Then there was the power of genetics. According to one researcher, "The genes of these hatchery fish may differ somewhat from those of the original lake trout populations, with the result that the hatchery fish may lack the behavioural and physical adaptions that allowed the original lake populations to thrive in the Great Lakes."[13] This theory was supported by research done in New York, where scientists found that trout introduced into Lake Ontario from nearby Seneca Lake survived almost twice as well as trout from other areas. Researchers also identified the possibility that accumulation of toxic chemicals in the bodies of the adult fish is passed to the embryo and prevents the successful hatching of eggs.

All of these problems underscored the fact that reintroduction was a tricky matter that could not alone lead to full restoration. But the work did show it was still possible, through a continued commitment to halting environmental abuses and through active research on the problems of rehabilitation, for lake trout to once again flourish in the waters of the Great Lakes.

Fish were not the only target for rehabilitation. Programs were also put in place to restore some of the other species of birds and animals that in years past were either eliminated from the region or came to exist only in diminished numbers. The states of Michigan and New York, and the province of Ontario, for example, initiated efforts to return wild turkeys to areas where they were once abundant. Similarly, martens, which had been extirpated from Michigan, were once again back prowling some of the state's forests, thanks to the transfer of animals captured in Algonquin Park, Ontario, where they had remained plentiful because of protection.

Many volunteers have joined in the work of helping to re-establish species that had suffered declines as a result of the activities of humans. Farmers, naturalist and sports clubs, and other interested individuals have set up nest boxes for bluebirds and wood ducks to enable these two colourful bird species to survive and regroup.

A VISION FOR THE FUTURE

Re-establishing lost species became, then, an important additional part of long-term environmental rehabilitation in the Great Lakes region. While there is no perfect formula for ecological rehabilitation, work has to begin with a definition of the desired result of the process. Towards this end, the authors of the Great Lakes Fishery Commission report compiled a list of what a rehabilitated Great Lakes ecosystem would need. They said that people in the region should be able to:

- drink water without fear of ingesting harmful viruses, bacteria, protozoa, and poisons.
- eat fish and waterfowl knowing that they are relatively uncontaminated by dangerous chemicals.

- harvest commercially valuable species of fish and furbearers in profitable quantities.
- swim without becoming infected by disease or soiled by waste film on the water surface.
- canoe or sail without encountering surface scums of wastes and offensive floatables.

It is an idealistic list, perhaps, and it seems likely that many a year will pass before the Great Lakes can be rehabilitated to the point where these visions of the future become anywhere close to reality. But the dedicated efforts of individuals and groups to stop the most flagrant abuses of the Great Lakes and rehabilitate some of its important species point the way ahead. These efforts strongly confirm public support for a healthy and thriving environment, and offer up a firm challenge for action over the next few decades.

TWELVE

Building New Foundations for Great Lakes Development

DR. JACK VALLENTYNE is a veteran of the war against pollution in the Great Lakes. By training he is a limnologist, a student of lakes, and his credentials are impressive. During the 1960s his experiments on the effect of phosphorus on the growth of algae were instrumental in convincing the governments of Canada and the United States to control wastes more effectively and to limit the level of phosphorus in detergents. After thirty years spent analysing lake life, he has over one hundred scientific publications to his credit and is the senior scientist with the Canada Department of Fisheries and Oceans. He is also the co-chairman of the Science Advisory Board of the International Joint Commission.

In the late 1980s, when I first met him, Vallentyne was working at the Canada Centre for Inland Waters. His office window, running the full length of the room, afforded a superb view of the Burlington ship channel and Hamilton Harbour. A lake freighter belonging to the Canada Steamship Company was slowly making its way from the lake to the steel mills that line the south shore of the harbour. The day's potentially clear blue skies were noticeably hazy from the gases pouring out of the smokestacks of the mills. Grey-winged gulls and smaller terns wheeled above the shimmering blue waters of the bay – waters heavily polluted with phenols, PCBs, and polyaromatic hydrocarbons (PAHS). When I commented on the scene, Vallentyne responded in a quiet, disgusted voice, "Yes, it is a reminder of what has gone wrong."

According to Vallentyne, we are intimately connected to our environment whether we like it or not – whether it's good for business or not. With a boyish grin he noted that in my lungs there is likely to be at least one molecule from the breath of every adult who has lived during the past three thousand years. The chloride ions that I am swallowing could have been almost anywhere previously: yesterday in a willow tree; centuries ago in the sweat on the back of a New World explorer; or even further back, in the waves of a prehistoric ocean.

Vallentyne's manner of presenting this message turned out to be highly unorthodox. As often as possible he would set aside the white lab coat of the scientist in favour of a backpack with a globe of the earth attached to it – a symbol for him of the connection between people and their environment. He had taken to carrying this globe, he said, whenever he was more than five miles from home. It went with him to the theatre, to scientific meetings, and on his travels around the world.

On the morning of my visit, Vallentyne had been wearing the globe in front of his favourite audience, school children. He said that by influencing people at a young age he hoped they would begin to see themselves as part of the ecosystem, which "is really just another word for home." Said Vallentyne, "If we treat our environment as a home, something we are a part of, it will have a profound effect on our actions." People will not then be thinking of their environment as something separate – they will be thinking of themselves as being in an ecosystem. As Vallentyne put it, this clearly calls for people to "think, talk, and act" in different ways. As I listened to him I realized that our failure to recognize ourselves as part of the ecosystem has allowed us to unwittingly change the system in ways that have destroyed its health.

THE DECAYED FOUNDATIONS

The dramatic transformation of the Great Lakes ecosystem began with the everyday approach that European settlers and their enterprises took to the land and the life it supported. As contemporary philosopher William Leiss put it, for Europeans "Nature was said to require the superintendence of man in order to function well, and this was understood as a thorough transformation of the natural environment, rather than mere occupation or nomadic passage."[1]

This transformation was driven by what we can now see as false beliefs. The first of those beliefs was that natural resources were inexhaustible, an attitude that led to greater and greater exploitation and utilization of greater and greater quantities. The very term "resources" entered into use in the sixteenth century and, as Christopher Vescy noted, indicated "a view of the environment as material to be exploited."[2]

From the early fur trade, which led to the depletion almost to the point of extinction of beavers and martens, to the timber trade, which eradicated the massive white pine forests, to the extinction of the passenger pigeon and the blue pike, the people failed to learn the lessons of nature, failed to recognize the limits that existed.

For there are indeed limits to the growth of trees, limits to the amount of productive agricultural land in a region, and limits to the number of fish that lakes can support. Through painful experience we have learned that the resources of the Great Lakes region are not inexhaustible. But, unfortunately, to this day most industrial and business enterprises continue to operate on the mistaken belief that we can forever increase our consumption of resources, including land, minerals, forests, and fish.

There have been other false beliefs as well. We assumed for the longest time that the environment could indefinitely absorb and assimilate the disposal of our society's unwanted byproducts. As the development of the Great Lakes region progressed, and the problems of human and animal waste disposal became apparent, we managed to ignore those problems, at least until the spread of disease by water contaminated by wastes forced the governments to ask the International Joint Commission to investigate the problem. The polluted state of the waters, as reported by the IJC in 1918, should have served as sufficient warning of the dangers of the dilution approach to pollution. Sadly, decision-makers ignored this early warning sign and officials introduced the "cheaper" alternative of chlorination. The dilution approach held fast.

The environmental destruction caused by the dumping of wastes has not gone away. Although governments and manufacturers have stopped discharging many wastes directly into water, they have simply transferred other wastes to land or the atmosphere. They have avoided long-term solutions in favour of short-term "out-of-sight-out-of-mind" approaches. Meanwhile, the types of wastes produced have increased both in number and in powers of destruction.

Yet another false belief rested on the idea that human actions and activities had a certain simplicity, a clear-cut direction that could be readily understood and predicted. But the Great Lakes ecosystem that sustains us is a complex matter. Actions in one part of it invariably have repercussions elsewhere. For the most part, human activity in the Great Lakes basin proceeded without recognition of the environmental impact. The initial clearing of the land by settlers unwittingly resulted in massive erosion of soils, siltation of fish spawning areas, and destructive alteration to the delicate hydrologic cycle. Human actions in later years continued to present unwanted ecological surprises. The construction of canals, for example, while greatly expanding shipping activity, allowed the destructive lamprey and alewife to get into the lakes. Similarly, the use of chemical pesticides initially reduced insect damage to crops, but some of the insects became resistant and the expanding use of chemicals poisoned the environment.

One final false belief: that environmental problems can be solved by quick-fix technological solutions. At best such solutions have proved to be shortlived and incomplete and at worst have caused major environmental destruction. Typical of this were recent suggestions to "control" lake levels and to divert water from the lakes to water-short areas in the United States. The lake levels naturally fluctuate in response to rainfall and other natural processes, creating a seasonal and yearly rise and fall of the lakes that is, for instance, ecologically necessary for many lakeside wetlands. Removing these natural rhythms would arrogantly impose on nature a human regularity to which it is unaccustomed.

Development in the Great Lakes region has for too long proceeded in a manner that has failed to account for potential ecological implications. All too often we have been surprised by the results of seemingly simple actions. Our survival in the region depends upon our ability to anticipate as much as possible the full consequences of our activities and to prevent developments that are going to cause damage to some part of the ecosystem. We can no longer afford to make substantial technological or artificial changes without a complete accounting of the repercussions. We can no longer afford to be surprised by the extent and intensity of the side-effects of our actions.

The future of the region depends on discarding the belief in the inexhaustible nature of natural resources, in the unlimited ability of the environment to assimilate our waste products, and in the simplicity of our actions. Throwing away these beliefs is the first step towards a healthy and sustainable future in the Great Lakes region.

ECOSYSTEM MANAGEMENT

No one person or group is responsible for the management of the Great Lakes region. Two federal governments, eight states, two provinces, countless municipal, county, and regional governments, a large number of private companies and organizations, and individuals all make decisions that affect the ecosystem – though not all have equal power in this. The collective actions of the thirty-seven million people who live within the basin and of their governments and organizations chart the course of the region's development.

During the last hundred years, the governments of Canada and the United States have developed and signed various treaties, agreements, and conventions in the hopes of better managing the ecosystem – or at least of controlling some of the damage. Most important of these was the Boundary Waters Treaty of 1909, and the Great Lakes Water Quality Agreements of 1972 and 1978. The 1978 agreement was amended in November 1987. The preamble to the new agreement emphasized "the need for strengthened efforts to address the continuing contamination of the Great Lakes Basin Ecosystem."[3]

Fred Brown, the president of Great Lakes United, reinforced the need for such an emphasis and praised the new amendments, but noted that the real test would be how well they were carried out. "These amendments have yet to remove one iota of toxic substance from the Great Lakes," he said.[4] More than a decade after the 1978 agreement and its emphasis on wise resource use, preventative measures for environmental problems, and a stronger effort towards institutional co-operation and co-ordination in decision-making, there remained little argument over the need for such an approach, but its implementation has been painfully slow.

Together, the existing agreements, conventions, and treaties have in one way or another helped slow the pace of degradation that has persisted in the Great Lakes region. They have not, however, led to an abandonment of the system of development and resource management that undermined the ecological health of the ecosystem in the first place. The long-term health and sustainability of the ecosystem still required new foundations for development.

MORE WITH LESS: WISE RESOURCE USE

Natural resources are the key to the continuing development of the Great Lakes region. A strategy of ecosystem management requires that resources be used carefully and that the maximum potential of each resource be realized. To do otherwise is to invite environmental ruin.

There is great potential for more effective use of resources. One Canadian study found that through energy conservation alone Canada could meet its future expanding energy needs to the year 2000. In her preface to the *Green Consumer Guide,* Canadian author Margaret Atwood lists just a few of the things that people can do to conserve resources:

> Dry your clothes on a rack: humidify your home and lower your hydro bill.... Start a compost heap.... Choose non-disposables: razors with real blades instead of the plastic chuck it out kind, fountain pens rather than toss-outs. Shop for organic veggies; do it using a shopping basket so you won't have to cart home all those annoying plastic bags that pile up under the sink.[5]

The motto to guide management of non-renewable resources in particular, and resources in general, must be "to do more with less." It is a simple philosophy, but one that could have tremendous repercussions for the health and sustainability of the ecosystem. Fisheries, forests, and land are all renewable resources that, if properly cared for and managed, can continually aid human well-being.

A step towards achieving this goal was made by fish managers grappling with the problem of sustaining stocks in the Great Lakes. In December 1980, U.S. federal and state fishery agencies and Canadian federal and provincial fishery agencies signed "A Joint Strategic Plan For the Management of Great Lakes Fisheries." The stated goal of the plan was "to secure fish communities based on foundations of stable self-sustaining stocks, supplemented by judicious plantings of hatchery-reared fish."[6]

As an essential element the plan recognized ongoing consultation and co-operation among sports and commercial fishers in developing a fisheries management plan. To protect the fish-spawning and nursery habitats and to address the problem of toxic chemicals, fisheries managers have to co-ordinate their activities with the IJC and other agencies responsible for water quality and land use.

In the forest industry, companies are supposed to replant cutover lands and to adopt improved regeneration techniques in the forest areas being harvested. All too frequently, however, forest companies are not leaving behind regenerated and healthy, productive forests after cutting. The failure to renew healthy forests runs counter to an ecosystem approach to resource use and will undermine ecologically and economically sustainable development. It is essential, too, that the forests of the region are not managed solely for the benefit of tree growth. Other uses of the forest – for wildlife habitat, tourism, hunting, hiking, birdwatching, and photography – must be respected. They also require different management approaches.

Naturalists in Ontario successfully argued the case for such consideration in a battle over one of the largest remaining Carolinian woods in the province. The naturalists wanted the woodlot left alone without human interference. "There are only a few places left in the county where a sense of grandeur of our original forest remains," wrote local naturalist Michael Bradstreet. The group argued that protection of these areas from even selective cutting was necessary to provide habitat for a number of rare and endangered plants and animals and to preserve undisturbed remnants of the original forests. Bradstreet contrasted the naturalists' desires with the current trends, where "forest cover has come to mean regenerated rows of pines, or young stands of hardwoods with the venerable trees removed."[7]

If we carefully control inappropriate uses of highly productive forest lands and develop techniques for regenerating forests that ensure ecosystem health as well as good tree production, we ensure our material well-being and maintain environmental health. We can also ensure employment opportunities in the forest industry today and in the future.

Similarly, farmlands, wetlands, and other productive and ecologically important areas also need protection. Wetlands, described by the U.S. Environmental Protection Agency as one of the most "valuable, and perhaps irreplaceable resources," have been lost at an alarming rate. Soil loss throughout the region has also been staggering. In fields under continual corn cultivation, soil losses have been calculated at 5.5 tons per acre on level land and up to 21 tons on land with a 10 per cent slope. Such losses of soil do not happen if good soil management practices are employed. Both wetlands and soil losses are generally irreversible and will in the long run undermine our economic and ecological well-being.

Using the resources of the region wisely ensures their potential for providing long-term benefits. In words that remain as relevant today as when they were first written, the U.S. National Academy of Sciences told President John F. Kennedy in 1963: "The objectives of our national policy in the management of our national resources should be rooted in our concern for our children and the generations to follow, and automatically guarantee to them, without loss of flexibility of action today, the widest possible range of choice."[8]

The continued survival and well-being of all who live in the Great Lakes region depend on the continued health of the natural resources that make up the ecosystem. The limits of the resource do not have to constrain our well-being and happiness. But these limits do define the exciting challenge we face in learning how to live our lives in a way that does not destroy the environment upon which we depend for our survival.

MANAGING WASTES: A PREVENTATIVE APPROACH

No practice in the Great Lakes region has had such damaging consequences to environmental health as the management or – more appropriately – mismanagement of wastes. Typhoid and cholera epidemics were two deadly consequences of this mismanagement, and the eutrophication of Lake Erie another. The most recent manifestations are the mountains of municipal trash looking for a home and the poisoning of the entire Great Lakes environment by minute quantities of toxic chemicals.

Wastes need not be damaging to either the environment or human health. Clearly, the quantities of waste generated in the late 1980s were excessive. In Denmark the average person produced 145 kilograms of waste a year, compared to an average of 1,540 kilograms per person in the United States. The average Canadian produced even more, at 2,270 kilograms.[9]

The decaying state of Lake Erie in the 1960s dispelled the popular belief in the environment's ability to absorb massive quantities of wastes without any harm being done. The lake could not absorb the huge volumes of sewage being dumped in it and eutrophication was the result. The subsequent efforts to build sewage-treatment plants and limit the amount of phosphorus in detergents proved that wastes could be effectively controlled given suitable approaches and resources. Such success, however, was not complete and left a need for further improvements in the control of municipal wastes. The ongoing problem of combined sewage outflows, for example, remained a blight on the efforts to control sewage pollution. Toxic chemical pollution presented even greater danger.

By the late 1980s toxic chemicals remained the single greatest threat to the health of the Great Lakes – a threat recognized by the governments in their 1978 Great Lakes Water Quality Agreement. The agreement's goals – "zero discharge" and to "virtually eliminate the input of persistent toxic substances" – reflected, on paper, the abandonment of old philosophies for managing chemicals. In reality, however, much of the legacy of "what you can't see can't hurt you" remained. Despite a pledge to do so, governments did not go on to enforce regulations to ensure zero discharge of toxic substances. Many corporations continued business as usual, and hundreds of pipes and smokestacks continued to emit toxic chemicals to the environment.

Direct industrial discharges, indirect industrial discharges through sewage-treatment plants, discharges to the atmosphere, and leaking dumps: all these continued to provide rich conduits for deadly toxic chemicals passing into the environment. The failure to manage these substances at their source made their control in the environment extremely difficult. As the 1985 Water Quality Board reported, "Past experience with such persistent toxic substances as mercury, DDT, PCBS, dieldrin, and mirex, has clearly demonstrated that once these persistent toxic chemicals get into the system, it is extremely difficult to remove them."[10]

The only prudent and ecologically sensible approach to dealing with toxic chemicals was to prevent them from reaching the environment, and the technologies and strategies for this prevention already existed and needed to be applied. The *Toronto Star*, for example, introduced an ink recycler into its operations in 1978. Within eight months the daily newspaper had completely eliminated the need for the disposal of toxic ink wastes and the purchase of $40,000 a year in new ink.

Other companies took up the challenge of reducing or eliminating their waste so that it did not need disposal in the lakes or in an inevitably leaking landfill. The 3M Company of St. Paul, Minnesota, initiated a company-wide program called Profit From Pollution Pays and proved at its numerous plants not only that significant reductions could occur in the amount of waste produced but also that these reductions can be both economical and environmentally beneficial. The company estimated that it had eliminated the need to disperse 97,500 tons of air pollutants, 10,500 tons of water pollutants, and 242,000 tons of sludge and solid waste annually.

These achievements indicated the tremendous potential of companies for developing and using innovative and creative methods to stop the environmental release of toxic chemicals at their source as well as for conducting their operations in environmentally sensitive ways. "A little inventiveness on behalf of business towards the problem of waste management will lead to a new dimension in the competitive spirit and will result in the development of new markets for waste materials and recycling processes," declared the Kingston, Ontario, Chamber of Commerce.[11]

The failure of companies to deal properly with their waste imposed an unnecessary burden on the environment and on people harmed by those chemicals. As Charles Caccia, former Canadian environment minister and a federal Member of Parliament, stated, "Pollution, no matter what part of the nation it affects or which industry produces it, is a cost imposed on the economy of the nation. It is a cost in terms of cleanup and it is a cost to human health and of subsequent health treatment."[12]

Governments did continue to make new commitments to control toxic chemicals. On May 19, 1986, the governors of the eight Great Lakes states announced that they were signing a Great Lakes Toxic Chemical Control Agreement, its purpose to "establish a framework for coordinated regional action in controlling toxic chemical pollutants entering the Great Lakes system."[13] They pledged themselves to a greater degree of information sharing about the hazards of toxic chemicals. Such a basin-wide agreement, noted the Michigan Department of Natural Resources, "allows us to protect the lakes based on boundaries of nature, not mere political boundaries."[14]

Only time would tell whether this additional agreement could indeed bring about action to strengthen the controls on toxic chemicals. What was apparent, however, was the need for a strengthened political will to reduce chemical contamination.

Environmental organizations and Great Lakes citizens have taken up the challenge of ensuring the development of that necessary political will. "Zero Discharge Now" was a consistent refrain of citizens at the biennial meeting of the International Joint Commission in Hamilton, Ont., in October 1989. On the first day of the two-day meeting a dozen members of the environmental group Greenpeace had paraded into the commission's luncheon carrying a coffin and wearing paper-mache masks of fish "covered with angry red tumors" and birds "with twisted, crossed beaks."[15] Joyce McLean, a Greenpeace member, delivered the keynote speech to the meeting – the first time the keynote had ever been delivered by an environmentalist. In her speech McLean noted that while the IJC had been discussing pollution of the Great Lakes since 1918, the two governments had gone on to ignore most of its recommendations.[16] According to the U.S. General Accounting Office, the IJC had made fifty-nine recommendations in 1978 but only twenty-nine had been implemented by 1989.[17] The outpouring of concern at the IJC meeting was so overwhelming that the commission had to schedule an additional day of hearings to accomodate all those who wanted to speak.

The Canadian Institute for Environmental Law and Policy (CIELAP) and the National Wildlife Federation also began a major research effort to develop model regulations and programs to achieve zero discharge of persistent toxic substances. In response to these and other efforts the U.S. Environmental Protection Agency initiated a program to develop uniform Great Lakes water quality criteria. In Ontario, the newly created Municipal Industrial Strategy for Abatement (MISA) planned to revise Ontario's water permit laws. Meanwhile, CIELAP and other Canadian environmental organizations were watching out to ensure that the new regulations both north and south of the border would reflect the ideals of the Great Lakes Water Quality Agreement. At the October 1989 IJC conference the coalition of Canadian and U.S. environmental groups called for a ten-year timetable for ending the discharge of pollution into the water. "We have to have clear goals," John Jackson of Great Lakes United told the meeting. "The risks are too high now. We cannot afford any more risks."[18] The citizens who spoke at the IJC conference noted in a media release: "The overwhelming rallying cry of the people of the Great Lakes is that there is *no more time to waste.* The time for action is now."[19] At the meeting John Jackson presented a three-stage plan for fighting the pollution problems:

1. A "toxic freeze" banning new polluters from putting up pipes or smokestacks in the region.
2. A crackdown on existing polluters when their smoke-discharge and sewer-discharge permits come up for renewal; they would be required to scale down their pollution.

3. An attack on non-point sources of pollution, such as runoff from city streets and farms where groundwater is loaded with pesticides.

Jackson said that consumers could help the third stage along by telling farmers and stores that they want pesticide-free food.

One continuing and important problem concerning toxic chemical management strategies has been the failure of governments to enforce regulations regarding the release of toxic chemicals. Helen Henrikson, a member of the Little Cataraqui Environmental Association, summarized the frustration of many Great Lakes citizens when she told a hearing on the Great Lakes Water Quality Agreement in August 1986, "We all recognize that the real problem is the lack of will on the part of the two parties to implement and enforce the terms of the Agreement."[20]

All the evidence indicated that political will indeed was a major factor in undermining efforts to improve the water quality of the Great Lakes. The 1989 report of the Great Lakes Water Quality Board, for example, showed that in Ontario 48 per cent of the major municipal wastes and over 58 per cent of industrial waste discharges to the Great Lakes did not meet the legal discharge requirements for at least one parameter. The enforcement in the United States is apparently equally lax. Complete assessment is difficult because the states do not provide adequate information to the IJC. The 1989 report presented data for only one state. Violations for discharges of phosphorus and BOD were 26 and 67 per cent respectively. The governmental reluctance to impose the necessary control threatens the future health of the Great Lakes ecosystem and its inhabitants.

Preventative approaches to environmental problems do not have to be restricted to waste management alone. New harbours, steel mills, or power plant projects can have deleterious consequences for the environment if they are built in improper locations or with inappropriate technology. Environmental reviews of such proposals afford at least a minimum of environmental protection by ensuring a positive impact on development and avoiding damage to the ecosystem.

CO-OPERATION AND CO-ORDINATION

An essential ingredient in an ecosystem approach to managing the Great Lakes basin is co-ordination and co-operation among various institutions throughout the region. The invasion of the lamprey into all the Great Lakes was a graphic reminder of the interconnectedness of the ecosystem and the people within it. Similarly, a decision by one government agency to introduce a new species of fish can have repercussions in far-flung parts of the region. We cannot ensure that a fish released in Canadian waters will swim only in the Canadian half of the lakes or that a fish released in Illinois will be content to live only in that state's waters. To make decisions that ignore implications elsewhere is foolhardy and at times criminal. An example of this was a remarkable episode involving officials of the Michigan Department of Natural Resources. One day officers from the department posted signs on a river warning fishers of high contaminant levels in fish; the next day other officers were busy stocking the river with sport fish.

A necessary component of proper management is to know not only what is happening in other parts of your own department, or industry, but also what is happening to the ecosystem – to be able to detect change, and to know whether policies and actions are working. For this to occur there must be a program in place to take the pulse of the ecosystem. The U.S. and Canadian governments attempted to do this by giving the International Joint Commission the responsibility for co-ordinating a program to measure and monitor the health of the aquatic ecosystem. In 1985 the IJC upgraded the system, called The Great Lakes International Surveillance Plan (GLISP), to better inform them about what was happening to the lakes. In its most recent report the Commission noted that the programs under GLISP are not being fully conducted by the parties involved, and the region needs a strengthened commitment to those programs. The region also needs other similar programs to assess the health of the entire ecosystem, including land and air – and especially a program to monitor toxic air contaminants.

An effective program of monitoring the ecosystem requires more than chemical tests on water or air. Watching for changes in the populations of birds, amphibians, and other wildlife can give early warning signals of stress in the ecosystem. The experience of the 1960s and early 1970s gave convincing proof of this need to monitor wildlife: during that period the reproductive failure among fish-eating birds warned of the damage to the environment being done by toxic chemicals.

One more recent sign of stress is the explosive spread of zebra mussels, first spotted in the Great Lakes in 1986. Scientists believe that the little shelled creatures, originally from the Black Sea area, arrived in North America as stowaways on foreign ships and were introduced to the water near Sarnia when the vessels emptied their ballast. By 1988 the mussels had established themselves to stay in Lake St. Clair, and then they spread to Lake Erie. In 1989 they reached Lake Ontario. The mussels get into water pipes, attaching themselves to the insides of the pipes with their biological "threads" and sticking on like glue. They grow larger and pile on top of one another until the water can't get through and the pipes corrode. The damage to municipal and industrial water pipes on both sides of the border in Lake St. Clair and Lake Erie was expected to be more than $5 billion in the 1990s. In the city of Monroe, Michigan, near Detroit, the mussels took control of the local water supply in December 1989, "forcing residents to close schools, stores and restaurants, cancel Christmas parties and give up showering and washing the dishes for three days."[21] The mussels had also choked off 40 per cent of the water intake in Tilbury in southwestern Ontario on Lake Erie. According to journalist David Israelson, the zebra mussels were "already doing about $200 million damage to the Great Lakes' commercial fishery and the ultimate potential of their threat to fish and wildlife is still unknown."[22]

In early March 1990 Ohio Senator John Glenn and Congressman Henry Nowak introduced a U.S. bill proposing a $40 million mussel control program, declaring the mussels to be "an ecological disaster of oil spill proportions." Again, the invasion of the zebra mussels called for international co-operation as well as local efforts to repair the damage already done to water pipes. By March 1990 the Great Lakes Fishery Commission was spearheading a team effort involving more than forty Canadian and U.S. officials from different agencies, working to find a way to get all ships to keep their ballast out of the Great Lakes.

There is an astonishing number of government organizations and interests with a stake in the management of the Great Lakes environment. For proper ecological management, all these agencies must co-ordinate their efforts and practise a degree of co-operation and consultation far beyond anything that exists at present. Decision-making based on such co-operation can be a creative and exciting process.

Such co-operative action is not without precedent in the region. The Boundary Waters Treaty of 1909 was remarkably farsighted in recognizing the need for international co-operation in the management of the lakes. More recently the Great Lakes governors and premiers signed two different charters making commitments to management of the water resources of the basin in an ecosystemic way. There are also less formal methods of co-operation and co-ordination. If resource managers and user groups in different jurisdictions worked to share their own information and experiences, this would undoubtedly lead to better management of the ecosystem. As Robin Lunn of the Bay of Quinte Environmental Group noted, "There are no borders to pollution and there should be no borders between the people who are working to clean it up."[23]

THE CHALLENGE OF HEALTHY AND SUSTAINABLE DEVELOPMENT

In July 1984, sixty Great Lakes region university scientists and researchers gathered at the Wingspread Conference Centre in Racine, Wisconsin, to discuss future development in the Great Lakes region. The focus of the conference was the challenge to create new ecologically and economical sustainable patterns of development.

On the second day of the conference, George Francis, a member of the Science Advisory Board to the IJC and a professor at the University of Waterloo, his university colleague Sally Lerner, and University of Michigan professor David Hale led the conference participants through an exercise called "futuring." Divided into workshop groups, participants had to let their imaginations run free to explore an ideal future vision of the Great Lakes region. After a few minutes spent getting over inhibitions about the exercise, ideas began to flow. The responses were somewhat predictable: water free from pollutants, public access to clean beaches, abundant opportunities to fish, plentiful parks and open space for hiking and camping, a healthy economy, and full employment. But the predictability of the responses did not disappoint the exercise organizers. "A strong and shared vision of a preferred future is necessary to mobilize the will and commitment to try and bring it about," they said.[24]

The overwhelming conclusion of participants at the conference – and a message increasingly heard throughout the Great Lakes region – was that the destruction of the environment is not the inevitable consequence of human activities. Harlan Hatcher, historian and Director of the Centre for the Great Lakes, emphasized this point:

> The message of Earth Day was that we must nourish the Earth if we expect the Earth to nourish us. We at the Centre for Great Lakes are fond of saying the same thing another way: the economic vitality and environmental protection of the Great Lakes are inextricably linked. To preserve one is to protect the other – now and for the future.[25]

Clean water, healthy and abundant natural resources, and human well-being can be compatible with a healthy economy. The challenge facing us in the Great Lakes region is to sustain economic and human health while protecting, preserving, and renewing the ecosystem that maintains us. This means placing the stress on sustainable development: living a lifestyle that perpetuates ourselves and our environment and struggling for political, economic, and social policies that place priorities on people and not on profits and false promises of progress. Sustainable development does not mean attempting to sustain and perpetuate our current foundations for development. It means building on new foundations that do not destroy the environment on which we depend.

We have already come a long way in our understanding of the ecological processes at play in the Great Lakes ecosystem. Our long-term survival depends on that ability to understand the implications of our actions on the natural world around us, and on our will to take action to protect our integrated world.

Notes

Notes for Chapter 1

1. International Reference Group on Great Lakes Pollution From Land Use Activities (PLUARG), *Environmental Management Strategy for the Great Lakes System: Final Report to the International Joint Commission* (Windsor, Ont., 1978), p.1.
2. Much of the information for this section comes from: David M. Gates, C.H.D. Clarke, and James T. Harris, "Wildlife in a Changing Environment," in Susan L. Flader (ed.), *The Great Lakes Forest* (Minneapolis: University of Minnesota Press, 1983); Glenda Daniel and Mark Reshkin, "Bedrock and Landscape," in *Decisions for the Great Lakes: A Project of Great Lakes Tomorrow* (Hiram, Ohio: Great Lakes Tomorrow; and Hammond, Indiana: Purdue University Calumet, 1982); R. Cole Harris (ed.), *Historical Atlas of Canada, Vol.1: From The Beginning to 1800* (Toronto: University of Toronto Press, 1987); and Walter M. Tovell, *The Great Lakes* (Toronto: Royal Ontario Museum, 1979).
3. Much of the information in this section comes from Gates, Clarke, and Harris, "Wildlife."
4. Clifford E. Ahlgren and Isabel F. Ahlgren, "The Human Impact on Northern Forest Ecosystems," in Flader (ed.), *Great Lakes Forest*, p.35.
5. Cited in Stanford H. Smith, "Pushed Toward Extinction: The Salmon and Trout," in John Rousmaniere (ed.), *The Enduring Great Lakes* (New York: W. W. Norton, 1979), p.36.
6. Fred Landon, *Lake Huron* (Indianapolis / New York: Bobbs-Merrill, 1944), p.102.
7. Ron Reid, "Wetlands," in *Decisions for the Great Lakes*, p.818; and Elizabeth A. Snell, *Wetland Distribution and Conversion in Southern Ontario*, Working Paper No. 48 (Ottawa: Inland Waters and Lands Directorate, Environment Canada, 1987).
8. Quoted in Martin Kaatz, "The Black Swamp: A Study in Historical Geography," reprinted from *Annals of the Association of American Geographers*, Vol. XLV, No 1 (March 1955).

9. Eugene F. Stoermer, "Bloom and Crash: Algae in the Lakes," in Rousmaniere (ed.), *Enduring Great Lakes,* p.14.
10. Information for this section comes from: Bruce G. Trigger, *The Huron: Farmers of the North* (New York: Holt, Rinehart and Winston, 1959); George T. Hunt, *The Wars of the Iroquois* (Madison: The University of Wisconsin Press, 1967); Harris (ed.), *Historical Atlas*; and *Environment Canada, A Guide to Great Lakes Water Use Map* (Ottawa, 1980).
11. Hunt, *Wars of the Iroquois,* p.125.
12. Landon, *Lake Huron,* p.102.
13. Quoted in Wilbur Jacobs, "Indians as Ecologists and other Environmental Themes in American Frontier History," in Christopher Vecsey and Robert W. Venables (eds.), *American Indian Environments: Ecological Issues in Native American History* (Syracuse: Syracuse University Press, 1980), p.49.
14. Mark Reshkin, "The Natural Setting," in *Decisions for Lake Michigan* (Hammond, Indiana: Purdue University Calumet, 1979), p.26.

Notes for Chapter 2

1. The following sources helped to provide information for this section: Harlan Hatcher, *The Great Lakes* (Oxford: Oxford University Press, 1944); Harris (ed.), *Historical Atlas*; William Ellis, *Land of the Inland Seas: The Historic and Beautiful Great Lakes Country* (Palo Alto, Cal.: American West Publishing Company, 1974); and "The American Lakes Series" (Indianapolis / New York: Bobbs-Merrill), a collection of books on each of the Great Lakes by various authors.
2. Jerry Sullivan, Wayland Swain and Edward Pleva, "The Impact of Human Occupation," in *Decisions for the Great Lakes,* p.88.
3. Cited in Landon, *Lake Huron,* p.17.
4. Cited in ibid., p.21.
5. Louis Hennepin, "Voyage of the Griffin," in Walter Havighurst (ed.), *The Great Lakes Reader* (New York: The Macmillan Company, 1966), p.30.
6. Quoted in Penny Petrone (ed.), *First People, First Voices* (Toronto: University of Toronto Press, 1983), pp.59-60.
7. Cited in J.W. Chafe and A.R.M. Lower, *Canada: A Nation* (Toronto: Longmans, Green and Company, 1948), p.205.
8. Ellis, *Land of the Inland Seas,* p.188.
9. Chafe and Lower, *Canada,* p.205.
10. Charles E. Twinning, "The Lumbering Frontier," in Flader (ed.), *Great Lakes Forest,* p.123.
11. Ibid., p.129.
12. Cited in Milo Quaife, *Lake Michigan* (Indianapolis / New York: Bobbs-Merrill, 1944), p.327.
13. Ellis, *Land of the Inland Seas,* p.179.
14. Harlan Hatcher, *Lake Erie* (Indianapolis / New York: Bobbs-Merrill, 1945), p.181.
15. Quaife, *Lake Michigan,* p.227.
16. Cited in Arthur Pound, *Lake Ontario* (Indianapolis / New York: Bobbs-Merrill, 1945).
17. Quoted in Quaife, *Lake Michigan.*

Notes for Chapter 3

1. Cited in Smith, "Pushed Toward Extinction," in Rousmaniere (ed.), *Enduring Great Lakes,* p.36.
2. Ellis, *Land of the Inland Seas,* p.195.
3. Quoted in Landon, *Lake Huron,* p.105.

4. Randall E. Rohe, "The Upper Great Lakes Lumber Era," in *Inland Seas*, Vol.40, No.1 (Spring 1984), pp.16-28.
5. A. H. Lawrie and Jerald F. Rahrer, *Lake Superior: A Case History of the Lake and its Fisheries*, Technical Report No. 19 (Ann Arbor, Michigan: Great Lakes Fishery Commission, 1973), pp.13-14.
6. Quoted in Twinning, "The Lumbering Frontier," in Flader (ed.), *Great Lakes Forest*, p.124.
7. Reverend Peter Pernin, *The Great Peshtigo Fire*, reprinted from *Wisconsin Magazine of History* by the State Historical Society of Wisconsin, 1971.
8. Ibid.
9. Quaife, *Lake Michigan*, p.296.
10. Quoted in Lawrie and Rahrer, *Lake Superior*, p.29.
11. Quoted in Karen Heiber, "An Encore for Atlantic Salmon," in *Landmarks*, Vol. 5, No. 4 (Fall 1987), p.8.
12. Landon, *Lake Huron*, pp.133-134.
13. Ibid., p.134.
14. J.L. Goodier, "The Nineteenth-Century Fisheries of the Hudson Bay Company Trading Posts on Lake Superior: A Biogeographical Study," in *Canadian Geographer*, Vol. XXVIII, No. 4 (1984).
15. Cited in Lawrie and Rahrer, *Lake Superior*, p.30.
16. H. A. Regier and W. L. Hartman, "Lake Erie's Fish Community: 150 Years of Cultural Stress," in *Science*, Vol. 180 (June 22, 1973), pp.1248-1255.
17. Quoted in University of Wisconsin Sea Grant, *The Fisheries of the Great Lakes: 1984-1986 Biennial Report* (Madison, Wisconsin, 1986), p.5.
18. John Van Oosten, quoted by Frank Egerton in *Overfishing or Pollution? Case History of a Controversy on the Great Lakes*, Technical Report No. 41 (Ann Arbor, Michigan: The Great Lakes Fishery Commission, 1985), p.5.
19. John Power, "The Comeback of King Salmon," in *Landmarks*, Summer 1985, p.19.
20. W.J. Christie, *A Review of the Changes in the Fish Species Composition of Lake Ontario*, Technical Report No. 23 (Ann Arbor, Michigan: The Great Lakes Fishery Commission, 1973), p.31.
21. Claude Allouez, "Mission on Lake Superior," in Havighurst (ed.), *Great Lakes Reader*, p.13.
22. Quoted in Petrone (ed.), *First People, First Voices*, p.58.
23. From O.W. Main, *The Canadian Nickel Industry: A Study in Market Control and Public Policy* (Toronto, 1955), quoted in Jamie Swift and DEC, *The Big Nickel: Inco at Home and Abroad* (Toronto: Between the Lines, 1977), p.115.
24. Quoted in Havighurst (ed.), *Great Lakes Reader*, p.271.

Notes for Chapter 4

1. See, for instance, the enthusiastic account of progress in R. McLaughlin, *The Heartland* (New York: Time-Life Library of America, 1967), p.93.
2. Quaife, *Lake Michigan*, pp.310-311.
3. Kaatz, "The Black Swamp."
4. International Joint Commission (IJC), *Final Report of the International Joint Commission on the Pollution of the Boundary Waters Reference* (Washington-Ottawa, 1918), p.7.
5. *The York Canadian Freeman*, cited in A. M. McCombie, "Changes in the Physical and Chemical Environment of the Laurentian Great Lakes," in *A Symposium on Introductions of Exotic Fish*, presented at the Twenty-first Meeting of the Canadian Committee of Freshwater Fisheries Research (Ottawa, January 1968), p.26.
6. IJC, *Final Report*, p. 51.

7. Ibid.
8. Quoted in *Detroit Free Press*, June 22, 1982; cited in IJC files.
9. *The News*, December 31, 1887, quoted in P. Rogerson, "Water Pollution Abatement in the Great Lakes: 1890-1978," M.A. thesis (Department of Geography, University of Toronto, 1984).
10. Sullivan, Swain and Pleva, "Impact of Human Occupation," p.93.
11. *Treaty Between the United States and Great Britain Relating to Boundary Waters, and Questions Arising Between Canada and the United States*, signed in Washington, January 11, 1909.
12. The letter is printed in its entirety in IJC, *Final Report.*
13. *Treaty Between the United States and Great Britain.*
14. Ibid., p.52.
15. Michael Ondaatje, *In the Skin of a Lion* (Toronto: Penquin Books, 1988), pp.105-106.
16. *Treaty Between the United States and Great Britain*, p.46.
17. Ibid., p.9.
18. Hatcher, *Lake Erie*, p.300.
19. Lee Niederinghaus Davis, *The Corporate Alchemists: Profit Takers and Problems Makers in the Chemical Industry* (New York: William Morrow and Company, 1984), pp.116-117.
20. Wayland R. Swain, "Great Lakes Research: Past, Present, and Future," in *Journal of Great Lakes Research*, Vol. 10, No. 2, p.100.
21. Ibid.
22. International Joint Commission (IJC), *Report of the International Joint Commission, United States and Canada on the Pollution of Boundary Waters* (Washington-Ottawa, 1951), p.20.
23. Ibid., p.165.
24. Ibid., p.165.
25. Ibid., p.77.
26. Ibid., p.21.
27. Ibid., p.178.
28. Frank Egerton, *Overfishing or Pollution? Case Histories of a Controversy on the Great Lakes*, Technical Report No. 41 (Ann Arbor, Michigan: The Great Lakes Fishery Commission, 1985), p.8.
29. Quoted in Gerard Bertrand, Jean Lang and John Ross, *The Green Bay Watershed*, Report No. 229 (Madison, Wisconsin: University of Wisconsin Sea Grant College Program, 1976), p.150.
30. Tom Kuchenberg, *Reflections in a Tarnished Mirror* (Sturgeon Bay, Wisconsin: Golden Glow Publishing, 1978), p.23.
31. Ibid.
32. E. J. Crossman, "Changes in the Canadian Freshwater Fish Fauna," in *A Symposium on Introductions of Exotic Fish*, Twenty-First Meeting of the Canadian Committee on Freshwater Fisheries Research (Ottawa, 1968), p.8.
33. Ibid.
34. Ibid.
35. Kuchenberg, *Reflections*, p.38.
36. U.S. Department of the Interior, *Water Pollution Problems of Lake Michigan and Tributaries*, 1968, p.43.
37. Smith, "Pushed Toward Extinction," p.40.
38. Howard Tanner, quoted in Kuchenberg, *Reflections*, p.72.
39. U.S. Department of the Interior, *Water Pollution Problems*, p.43.
40. Interview with Don Misner, Port Dover, Ont., Jan.15, 1988.

41. Quoted in Kuchenberg, *Reflections*, p. 58.
42. Kuchenberg, *Reflections*.

Notes for Chapter 5

1. Quoted in McLaughlin, *The Heartland*, p.93.
2. Ibid.
3. International Joint Commission (IJC), *Pollution of Lake Erie, Lake Ontario and the International Section of the St. Lawrence River* (Ottawa, 1970), p.14.
4. Alan Edmonds, "Death of a Great Lake," in *Maclean's*, November 1, 1965, p.28.
5. "Time for a Transfusion," *Time*, August 20, 1965.
6. "Great Lakes: The Dead Sea," *Newsweek*, April 12, 1965. p.33.
7. Edmonds, "Death of a Great Lake," p.28.
8. Marlette E. Swenson, "Great Lakes or Great Sewers?," in *Canadian Business*, August 1966, p.22.
9. Interview with Don Misner, Port Dover, Ont., January 15, 1988.
10. Edmonds, "Death of A Great Lake," p.29.
11. Kuchenberg, *Reflections*.
12. Green Bay Press Gazette, *Green Bay: Portrait of a Waterway*, collection of reprinted articles (Madison, Wisconsin: University of Wisconsin Sea Grant, 1979), p.10.
13. Casey Burko, "View From Other Vantage Points," in EPA Journal, Vol. 11, No. 2 (March 1985), p.19.

Notes for Chapter 6

1. Quoted in Edmonds, "Death of a Great Lake," p.43.
2. Quoted in *Time*, August 20, 1965.
3. Barry Commoner, *The Closing Circle* (New York: Bantam Books, 1971), p.108.
4. Quoted in Swenson, "Great Lakes or Great Sewers?," p.23.
5. International Joint Comission (IJC), *Interim Report of the International Joint Commission on Pollution of Lake Erie, Lake Ontario and the International Section of the St. Lawrence River* (Washington-Ottawa, December 1965).
6. Interview with Dr. Jack Vallentyne, Burlington, Ont., June 12, 1986.
7. IJC, *Interim Report*.
8. United States Department of the Interior, Federal Water Pollution Control Administration, *Water Pollution Problems of Lake Michigan and Tributaries* (Chicago, 1968), p.24.
9. IJC, *Pollution*, p.21.
10. Ibid., p.67.
11. Statement of Mrs. Carol M. Kaltwasser, Housewives to End Pollution, at U.S. Department of the Interior, Federal Water Quality Administration, *Conference on the Matter of Pollution of Lake Erie and its Tributaries – Indiana – Michigan – New York – Ohio*, Fifth Session, Detroit, June 3-4, 1970, p.417.
12. Quoted in Wendy Wriston Adamson, *Saving Lake Superior: A Story of Environmental Action* (Minneapolis: Dillon Press, 1974), p.42.
13. Quoted in ibid.
14. Statement to IJC Hearing on Pollution of Upper Great Lakes and Land Use Activity, Duluth, Minnesota, December 7, 1972, pp.249-250.
15. Robert Peterson, Financial Secretary of the UAW Minnesota State Community Action Program Council, in a statement to IJC Hearing on Pollution of Upper Great Lakes and Land Use Activity, Duluth, Minnesota, December 7, 1972.
16. Quoted in ibid., p.62.
17. Commoner, *Closing Circle*, p.1.

18. Quoted in Lynton Caldwell, "U.S. Institutions," in *Decisions for the Great Lakes,* p.130.
19. United States, *National Environmental Policy Act of 1969* (NEPA), Public Law 9-190.
20. Richard B. Bilder, "Controlling Great Lakes Pollution: A Study in United States-Canadian Environmental Cooperation," in *Michigan Law Review,* Vol. 70 (1971-72), pp.478-479.
21. United States and Canada, *Great Lakes Water Quality Agreement,* April 15, 1972.
22. Canada, Office of the Prime Minister, *Press Release,* April 15, 1972.
23. Quoted in International Joint Commission (IJC), *Third Annual Report: Great Lakes Water Quality* (Ottawa-Washington, December 1975), p.6.
24. Quoted in McLaughlin, *The Heartland,* p.6.
25. IJC, *Pollution,* pp.89-90.
26. International Joint Commission (IJC), *Report on Great Lakes Water Quality for 1972,* p.12.
27. Canadian Detergent Industry, "Phosphates in Detergents: The Facts," news release, undated, p.2.
28. M. Bates, in a statement to IJC Hearing on Pollution of the Upper Great Lakes and Land Use Activity, Thunder Bay, Ontario, December 5, 1972.
29. IJC, *Report on Great Lakes Water Quality for 1972,* p.2.
30. IJC, *Third Annual Report,* p.14.
31. International Joint Commission (IJC), *Fourth Annual Report, Great Lakes Water Quality* (Ottawa-Washington, September 1976).
32. Tony Lang, "Great Lakes Born Again?," in *The Inquirer Magazine,* October 24, 1976, pp.30-44.

Notes for Chapter 7

1. Much of the information for this section comes from the following sources: Kuchenberg, *Reflections*; and Bernard R. Smith, J. James Tibbies and B. G. H. Johnson; *Control of the Sea Lamprey (Petromyzon Marinua) in Lake Superior, 1953-1970,* Technical Report No. 26 (Ann Arbor, Michigan: Great Lakes Fishery Commission, 1974).
2. Kuchenberg, *Reflections,* p.61.
3. Don Whitehead, *The Dow Story* (New York / Toronto: McGraw-Hill Book Company, 1968), p.240.
4. Kuchenberg, *Reflections,* pp.66-67.
5. Ibid., p.67.
6. M.R. Greenwood, *1968 State-Federal Lake Michigan Alewife Dieoff Control Investigation: Final Report of Field Operations Officer to the State-Federal Alewife Technical Coordinating Committee* (Washington: U.S. Department of Interior, May 1970), p.2.
7. Cited in Kuchenberg, *Reflections,* p.72.
8. Wayne H. Tody and Howard A. Tanner, *Coho Salmon for the Great Lakes* (State of Michigan Department of Conservation, Fish Division, February 1966), p.1.
9. Ibid.
10. Dr. Wayne Tody, quoted in Frank Mainville, "A New Type of Fishing," in Rousmaniere (ed.), *Enduring Great Lakes,* p.58.
11. Russell McKee, in *Michigan Conservation,* November-December 1967.
12. Power, "Comeback of King Salmon," p.20.
13. Jim Wood, "What the Great Lakes Mean to Sportsmen," in *Great Lakes United Newsletter,* Summer 1987, p.5.
14. Frank Prothero, *The Great Lakes Fishermen,* May 1981, p.20.

15. James Kitchell, "Keeping Score: The Great Lakes Predator – Prey Game in the Future of Great Lakes Resources," in *A Report on the Activities of the University of Wisconsin Sea Grant Institute 1982-1984* (Madison, Wisconsin: University of Wisconsin, 1984), p.22.
16. Cited in "Chinook Salmon Declining in Lake Michigan," *Seiche* (Minnesota Sea Grant), June 1989.

Notes for Chapter 8

1. Rachel Carson, *Silent Spring* (Boston: Houghton Mifflin Company, 1973), p.20.
2. Sergej Postupalsky, "The Bald Eagle Returns," in Rousmaniere (ed.), *Enduring Great Lakes*, pp.76-77.
3. Michael Gilbertson, "A Great Lakes Tragedy," in *Nature Canada*, January-March 1975.
4. Interview with Dr. Doug Hallett, Acton, Ont., March 20, 1986.
5. Leo Bernier, Ontario government cabinet minister, quoted in Ross Howard, *Poisons in Public* (Toronto: James Lorimer and Company, 1980), p.30.
6. IJC, *Report on Great Lakes Water Quality for 1972*, p.15.
7. Interview with Don Misner, Port Dover, Ont., January 15, 1988.
8. Great Lakes Water Quality Board, *Fourth Annual Report to the International Joint Commission* (Windsor, Ont., July 1976).
9. Great Lakes Water Quality Board, *Sixth Annual Report to the International Joint Commission* (Windsor, Ont., July 1978), p.3.
10. Tom Muir and Anne Sudar, *Toxic Chemicals in the Great Lakes Basin Ecosystem: Some Observations* (Ottawa: Environment Canada, 1987), p.7.
11. U. S. Department of the Interior, *Water Pollution Problems of Lake Michigan and Tributaries* (Chicago, 1968), p.44.
12. Quoted in T. Whillans, "Fish Community Transformation in Three Bays Within the Lower Great Lakes," M.A. thesis (Department of Geography, University of Toronto, 1977), p.158.
13. John E. Carroll, *Environmental Diplomacy* (Ann Arbor: The University of Michigan Press, 1983), p.134.
14. Davis, *Corporate Alchemists*, p.17.
15. Denis Konasewich, "The Most Difficult Problem: Toxins and the Great Lakes," in Rousmaniere (ed.), *Enduring Great Lakes*, pp.68-69.
16. John Quarles, *Cleaning Up America: An Insider's View of the Environmental Protection Agency* (Boston: Houghton Mifflin Company, 1976), p.174.
17. Great Lakes Water Quality Board, *Fourth Annual Report*, p.3.
18. Honourable Don Jamieson, "Notes for a statement by the Secretary of State for External Affairs at the signing ceremony of the Great Lakes Water Quality Agreement," November 22, 1978 (Ottawa), p.1.
19. Canada, Department of External Affairs, "Communique," November 22, 1978.
20. Ibid.
21. Canada and United States, *Great Lakes Water Quality Agreement of 1978*, signed at Ottawa, November 22, 1978.
22. Ibid., p.4.
23. Canada, Department of External Affairs, "Communique."
24. Great Lakes Water Quality Board, *Seventh Annual Report to the International Joint Commission* (Windsor, Ont., July 1979), p.4.
25. Ibid., p.3.
26. Great Lakes Water Quality Board, *Great Lakes Water Quality Board 1978 Annual Report* (Windsor, Ont., July 1979).

Notes for Chapter 9

1. International Joint Commission (IJC), *First Biennial Report Under the Great Lakes Water Quality Agreement of 1978* (Windsor, Washington, and Ottawa, June 1982), p.3.
2. Interview with Don Misner, Port Dover, Ont., January 15, 1988.
3. Noel Burns, *Erie: The Lake That Survived* (Totawa, New Jersey: Rowman and Allenheid, 1985).
4. Non-point Source Control Taskforce of the Water Quality Board of the International Joint Commission, *Non-point Source Pollution Abatement in the Great Lakes Basin: An Overview of Post-PLUARG Development* (Windsor, August 1983), p.2.
5. Ibid., p.5.
6. National Research Council of the United States and the Royal Society of Canada, *The Great Lakes Water Quality Agreement: An Evolving Instrument for Ecosystem Management* (Washington: National Academy Press, 1985), p.44.
7. Ibid., p.35.
8. Great Lakes Water Quality Board, *1987 Report on Great Lakes Water Quality: Report to the International Joint Commission* (Windsor, November 1987), p.78.
9. International Joint Commission (IJC), *70 Years of Accomplishment: Report for the Years 1978-79.*
10. Great Lakes Water Quality Board, *1987 Report*, p.7.
11. International Joint Commission (IJC), *Water Quality of the Upper Great Lakes* (Windsor, Washington, and Ottawa, May 1979), p.3.
12. Toxic Substances Committee of the Great Lakes Water Quality Board, *Toxic Substances Control Programs in the Great Lakes Basin* (undated), p.5.
13. Comptroller General of the United States, *A More Comprehensive Approach is Needed to Clean Up the Great Lakes* (Washington: U.S. General Accounting Office, May 1982), p.33.
14. National Research Council and Royal Society, *Great Lakes Water Quality*, p.48.
15. International Joint Commission (IJC), *Second Biennial Report Under the Great Lakes Water Quality Agreement of 1978* (Windsor, Washington, and Ottawa, 1984).
16. Dr. Beverly Paigen, "Health Hazards of Love Canal," in *Testimony presented to the House Subcommittee on Oversight and Investigations* (Washington, March 21, 1979).
17. Ann Hillis, "Inside 'Love Canal' Looking Out," in *1980 Annual Report of the Ecumenical Taskforce* (Niagara Falls, N.Y., 1980), p.47.
18. New York Public Interest Research Group (NYPIRG), *The Ravaged River: Toxic Chemicals in the Niagara River*, 1981, p.1.
19. The Niagara River Toxics Committee, *Report of the Niagara River Toxics Committee: Summary and Recommendations*, 1984.
20. Environment Canada, *A Layman's Guide to the Niagara River Toxics Committee Report: The Canadian Position*, unpublished draft, June 12, 1984, p.10.
21. Niagara River Toxics Committee, *Report*, p.14.
22. Laurie Montour, *Brief to the Great Lakes Water Quality Citizens' Hearing*, Sarnia, Ont., October 9, 1986.
23. Quoted in Jock Ferguson, "Probe Ordered Into Toxic Waste in Sarnia," in *The Globe and Mail*, November 15, 1985.
24. Quoted in Great Lakes United Water Quality Taskforce, *Unfulfilled Promises* (Buffalo, 1987), p.29.
25. U.S. Environmental Protection Agency, *Five Year Program Strategy for Great Lakes National Program Office 1986-1990* (Chicago: U.S. EPA, 1985), p.13.
26. Quoted in Great Lakes United Water Quality Taskforce, *Unfulfilled Promises*, p.21.

27. Interview with Dr. John Black, Buffalo, N.Y., February 3, 1986.
28. P. Bauman, "Cancer in Wild Freshwater Fish Populations with Emphasis on the Great Lakes," in *Journal of Great Lakes Research*, Vol. 10, No. 3 (1984), pp.251-253.
29. Quoted in Jon Brinkman, "Population Growth and Toxic Contamination of the Great Lakes," in TEF data, February 1985, p.4.
30. Thomas Erdmans, Richter County Museum of Natural History, *Presentation to Citizens' Review of the International Great Lakes Water Quality Agreement* (Green Bay, Wisconsin, July 14, 1986).
31. National Research Council and Royal Society, *Great Lakes Water Quality*, p.58.
32. "Lake toxins called health risk," *Toronto Star*, October 11, 1989, p.A1; see The Conservation Foundation and Institute for Research on Public Policy, *Great Lakes, Great Legacy?* (Baltimore, Maryland, 1989).
33. City of Toronto Department of Public Health, *Toronto's Drinking Water* (April 1984), p.6.
34. Dr. Sandy Macpherson, Letter to the Board of Health, Toronto, April 5, 1984.
35. The Conservation Foundation and Institute for Research on Public Policy, *Great Lakes, Great Legacy?*, p.21.
36. Ibid.
37. National Research Council and Royal Society, *Great Lakes Water Quality*, p.56.
38. L. Keith Bulen, "Toxics: Today's Great Lakes Challenge," in EPA Journal, March 1985, p.15.

Notes for Chapter 10

1. Telephone interview with Lee Botts, Chicago, June 10, 1986; see also Harold Henderson, "This Woman Wants to Work in Your Sewers," in *Reader* (Chicago Free Press Weekly), February 21, 1986.
2. Lynton Caldwell, "U.S. Environmental Law," in *Decisions for the Great Lakes*, p.114.
3. This quote and those following are from an interview with Sister Margeen Hoffman, Niagara Falls, N.Y., February 3, 1986.
4. Interview with Rick Findlay, Environment Canada, Toronto, June 13, 1986.
5. Bill Van Gaal, Canadian Auto Workers Environment Committee, "Statement to the Citizens' Review of the Great Lakes Water Quality Agreement," Toronto, October 23, 1986.
6. Quoted in Elaine Jaques, "Evolution of a Stewardship Group: A Case History," in *Seasons*, Autumn 1987, p.49.
7. Charlotte J. Read, "Saving the Indiana Dunes," in *Decisions for the Great Lakes*, p.292.
8. Great Lakes United brochure, undated.
9. Interview with David Miller, Buffalo, February 3, 1986.
10. Great Lakes United, *The Great Lakes Congressional Agenda* (Buffalo, Fall 1985), p.4.
11. Great Lakes United, *Great Lakes Resolution Summaries, 1983-1985* (Buffalo).
12. Letter to John Mylod, Hudson River Sloop Clearwater, from David Miller, Great Lakes United, Buffalo, January 15, 1986.
13. "Great Lakes United Wins Water Quality Suit," in *The Great Lakes United*, Vol. 1, No. 2 (July 1986).
14. Great Lakes United Water Quality Taskforce, *Unfulfilled Promises*, Preface.
15. Quoted in "Daniel Heads Federation," in *Lake Michigan Monitor*, Summer 1987, p.5.

Notes for Chapter 11

1. Interview with Dr Henry Regier of the University of Toronto, Toronto, January 11, 1987.
2. George R. Francis and Henry A. Regier, University of Waterloo and University of Toronto, "Proposal to Establish a Reference Group on Great Lakes Rehabilitation and Restoration," paper submitted for endorsement to the Great Lakes Fishery Commission (Sault Ste. Marie, Ont., June 1977), p.iv.
3. George R. Francis, John J. Magnuson, Henry A. Regier, and David R. Talhelm, *Rehabilitating Great Lakes Ecosystems,* Technical Report No. 37 (Ann Arbor, Michigan: Great Lakes Fishery Commission, 1979).
4. IJC, *Second Biennial Report,* p.7.
5. The information in this section comes primarily from: Hallett J. Harris, David R. Talhelm, John J. Magnuson, and Anne M. Forbes, *Green Bay in the Future: A Rehabilitative Prospectus,* Technical Report No. 38 (Ann Arbor, Michigan: Great Lakes Fishery Commission, 1982); Gerard Bertrand, Jean Lang and John Ross, *The Green Bay Watershed,* Technical Report No. 229 (Madison, Wisconsin: University of Wisconsin Sea Grant College Program, 1976); and telephone interview with Dr. Hallett (Bud) Harris, Green Bay, Wisconsin, July 8, 1986.
6. Rebecca Leighton, "Developing a Remedial Action Plan: The Green Bay Experience," in *The Great Lakes United,* Fall 1986, p.5.
7. Interview with Dr. Hallett (Bud) Harris, University of Wisconsin, Green Bay, Wisconsin, September 1987.
8. The Hamilton Naturalists' Club, Letter to all Stakeholders, Hamilton, Ont., November 24, 1985.
9. Interview with Gil Simmons, Hamilton, Ont., July 7, 1987.
10. Ad Hoc Group, "Rehabilitate Black Creek and Redevelop Toronto Area Waters," unpublished paper (Toronto, April 2, 1986).
11. Ibid., p.4.
12. Quoted in Wayne Munton, "They Call It CEE-FIP," in *Landmarks,* Winter 1986, p.6.
13. Susan Peterson, "A Reprieve for the Lake Trout," in *Outdoor America,* Spring Issue (1984), p.17.

Notes for Chapter 12

1. William Leiss, *The Domination of Nature* (Boston: Beacon Press, 1972), p.74.
2. Christopher Vecsey, "American Indian Environmental Relations," in Christopher Vecsey and Robert W. Venables (eds.), *American Indian Environments: Ecological Issues in Native American History* (Syracuse: Syracuse University Press, 1980), p.33.
3. The Government of the United States and the Government of Canada, *Protocol Amending the 1978 Agreement Between the United States of America and Canada on Great Lakes Water Quality, As Amended on October 16, 1983,* November 1987.
4. Quoted in news release issued by Great Lakes United, Buffalo, November 18, 1987.
5. Margaret Atwood, "Preface," in *The Canadian Green Consumer Guide: How You Can Help,* prepared by the Pollution Probe Foundation in consultation with Warner Troyer and Glenys Moss (Toronto: McClelland and Stewart, 1989), p.3.
6. Great Lakes Fishery Commission, *A Joint Strategic Plan for Management of Great Lakes Fisheries* (Ann Arbor, Michigan, December 1980).
7. Michael Bradstreet, "Backus Woods: One Man's Carolinian Canada," in *Seasons,* Summer 1985, p.19.
8. Quoted in Leonard Dworsky and David Allee, "Overview and U.S. Perspective," in *Decisions for the Great Lakes,* p.182.
9. Great Lakes Water Quality Board, *1987 Report on Great Lakes Water Quality* (Windsor, 1987).

10. Great Lakes Water Quality Board, *1985 Report on Great Lakes Water Quality* (Windsor, June 1985), p.1.
11. Kingston Chamber of Commerce, *Presentation to a Citizens' Review of the International Great Lakes Water Quality Agreement* (Kingston, Ont., August 19, 1986).
12. Charles Caccia, "Water: Learning From A Bitter Lesson," presentation to the Conference on Niagara River: The Next Ten Years (St. Catherines, Ont., Brock University, March 8, 1985).
13. Council of Great Lakes Governors, *The Great Lakes Toxic Substances Control Agreement* (Mackinac Island, Michigan, May 21, 1986).
14. Office of the Great Lakes, Michigan DNR, *The State of the Great Lakes: Annual Report for 1986* (Lansing, Michigan).
15. "Great Lakes sickness now threatens us, scientists say," *Toronto Star,* October 15, 1989, p.B1.
16. "Lake toxins ignored, forum told," *Toronto Star,* October 13, 1989, p.A10.
17. "Great Lakes sickness now threatens us, scientists say," *Toronto Star,* October 15, 1989, p.B1.
18. "Deadline urged for cleanup of Great Lakes," *Toronto Star,* October 14, 1989.
19. Great Lakes No Time To Waste Coalition, "Media Release" (Hamilton, Ontario, Oct. 12, 1989).
20. Helen Henrikson, *Presentation to a Citizens' Review.*
21. David Israelson, "Onslaught of mussels creating lake havoc 'worse than oil spill'," *Toronto Star,* March 12, 1990, p.A1,A14. See also "Zebra Mussel Infestation of the Great Lakes and the Effects on Water Intake Systems," U.S. EPA Fact Sheet (Chicago, September 1989); and Fred Snyder, "Zebra Mussels in Lake Erie: The Invasion and Implications," in *Twine Line* (Columbus, Ohio: Ohio Sea Grant College Program, Ohio State University, December 1989), pp.3-5.
22. Israelson, "Onslaught of mussels."
23. Robin Lunn, *Presentation to a Citizens' Review.*
24. Conference on Sustainable Redevelopment for the Future of the Great Lakes Region, "Conference Kit" (Racine, Wisconsin, Wingspread Centre, July 22-24, 1984).
25. Harlan Hatcher, "Private-Public Teamwork Needed," editorial in *The Great Lakes Reporter,* Vol. 2, No. 3 (May / June 1985), p.2.

Index

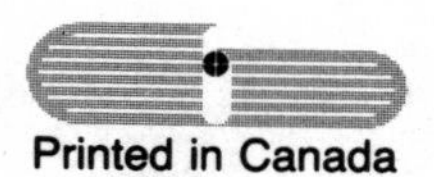
Printed in Canada